T0332422

Tomorrow's Systems Engineering

This book looks at systems engineering now and comments on the future. It notes the signs of deepening our understanding of the field which includes, digital engineering, interactive model-based systems, decision support frameworks, and points to a grand unified theory. The book also suggests how the systems engineer can be a better designer and architect.

Offering commentaries regarding how the field of systems engineering might evolve over the next couple of decades, *Tomorrow's Systems Engineering: Commentaries on the Profession* looks at the potential opportunities that might lie ahead rather than making predictions for the future of the field. The book allows the reader to prepare for the future in terms of technical interest as well as competitiveness and suggests opportunities that could be significant and useful for planning actions in the careers of future systems engineers. Discussions of improvements in how we develop and use software that can help to facilitate and protect overall IT capability within the system design and system architecture are also included.

This book is for systems engineers and software engineers who wish to think now about the directions the field might take in the next two decades.

Tomorrow's Systems Engineering
Commentaries on the Profession

Howard Eisner

CRC Press
Taylor & Francis Group
Boca Raton London New York

CRC Press is an imprint of the
Taylor & Francis Group, an **informa** business

First edition published 2023
by CRC Press
6000 Broken Sound Parkway NW, Suite 300, Boca Raton, FL 33487-2742

and by CRC Press
4 Park Square, Milton Park, Abingdon, Oxon, OX14 4RN

CRC Press is an imprint of Taylor & Francis Group, LLC

© 2023 Taylor & Francis Group, LLC

ISBN: 978-1-032-21622-5 (hbk)
ISBN: 978-1-032-21623-2 (pbk)
ISBN: 978-1-003-26927-4 (ebk)

DOI: 10.1201/9781003269274

Typeset in Times
by KnowledgeWorks Global Ltd.

This book is dedicated to:

- The forward-looking Fellows of INCOSE
- My wife June Linowitz, who knows how to live in the present and be mindful of the future
- My children Oren and Susan who have helped me both with my present and my future
- My five grandchildren, Jake, Gabe, Becca, Zachary, and Ben, who are all in their 20s, and who have assured their futures by their education achievements

Contents

Preface

This treatise looks briefly at the present in systems engineering and comments upon a possible future. It is distinctly not a prediction regarding the future. Rather, it contains conjectures about the future and in that sense might push us toward desirable scenarios in a decade or two. We note, with respect to the present, that there are signs of deepening our understanding of systems engineering. These include systems engineering as a discipline and pointers toward a grand unified theory. At the same time, we look at the systems engineer and how he or she might become better designers with higher fidelity and productivity. So hang on to your chair and stay with some new commentaries that will challenge you to "think outside the box". Staying inside the box usually doesn't lead to progress. Outside the box might lead to some new advances in some new as well as old domains.

Howard Eisner
Bethesda, Maryland
The George Washingon University

Author Biography

Howard Eisner spent 30 years in industry and 24 years in academia. In the former, he was a research engineer, manager, and executive at ORI, Inc. and the Atlantic Research Corporation and President of two high-tech companies: Intercon Systems and the Atlantic Research Services Company. In academia, he was professor of engineering management and a distinguished research professor in the engineering school of the George Washington University (GWU). At GWU, he taught courses in two departments, dealing with systems engineering, technical enterprises, project management, modulation and noise, and information theory.

He has written 12 books that relate to engineering, systems, and management. He has also given many lectures and tutorials to professional societies (such as INCOSE – International Council on Systems Engineering), government agencies (such as the Department of Defesnse (DoD), National Aeronautics and Space Adminisytration (NASA), and DOT), and the Osher Lifelong Learning Institute (OLLI). In 1994, he was given the outstanding achievement award from the GWU Engineering Alumni.

Dr. Eisner is a life fellow of the The Institute of Electrical and Electronics Engineers (IEEE) and a fellow of INCOSE and the New York Academy of Sciences. He is a member of Tau Beta Pi, Eta Kappa Nu, Sigma Xi, and Omega Rho, various honor and research societies. He received a bachelor's degree (BEE) from the City College of New York (1957), an MS in electrical engineering from Columbia University (1958), and a Doctor of Science degree from the George Washington University (1966).

Since 2013, he has served as Professor Emeritus of Engineering Management and a Distinguished Research Professor at the George Washington University.

OTHER BOOKS FROM THIS AUTHOR

- *Advanced Algebra*
- *Computer-Aided Systems Engineering*
- *Reengineering Yourself and Your Company*

- *Managing Complex Systems – Thinking Outside the Box*
- *Essentials of Project and Systems Engineering Management*
- *Topics in Systems*
- *Systems Engineering – Building Successful Systems*
- *Thinking – A Guide to Systems Engineering Problem Solving*
- *Systems Architecting – Methods and Examples*
- *Systems Engineering – 50 Lessons Learned*
- *What Makes a Successful Systems Engineer?*
- *Problem Solving – Leaning on New Thinking Skills*
- *Cost-Effectiveness Analysis – A Systems Engineering Perspective*

Toward a Future Environment

1

The field of systems engineering is currently under some stress, and perhaps is at a turning point. This introductory chapter examines some of the stresses of today and where they might lead. Tomorrow's field of systems engineering will be, of course, dependent upon today's environment. Some areas will move forward progressively and with positive effects and some otherwise. So let us start with a reiteration of this author's articulation of force fields from his book published in the year 2000 [1].

FORCE FIELDS 20 YEARS AGO

Some 20 years ago, this author identified several force fields that are relevant to look at here. These fields are:

1. The information age
2. Speed and responsiveness
3. Competition
4. New work patterns and environments
5. Loyalties and leverage

The Information Age

We are no less dependent today than we were 20 years ago. Indeed, the age has grown and intensified, especially since essentially all companies need information systems and must be protected against viruses and intrusions of various types. Looking forward, we see a deepening of the information age, with new problems to contend with and solve.

DOI: 10.1201/9781003269274-1

Speed and Responsiveness

The need for both has also intensified, led in industry by Amazon, for better or worse. Now that we can get faster responses, we are demanding such. It doesn't get much faster than packages sent in the morning and received by the afternoon of the same day.

Competition

It's at least as competitive today as 20 years ago, exacerbated by the COVID-19 virus. Various enterprises have had to be more competitive in order to survive. And the heavy duty retailers (like Walmart and Amazon) continually demand competitive prices on just about everything.

New Work Patterns and Environments

The COVID-19 virus has pushed us into more work from home as have the natural competitive forces. No one seems to be complaining about being at home to get work done, and saving gas monies while not fighting in traffic.

Loyalties and Leverage

These seem about the same, although many people have expressed the notion that they would change jobs in an instant if they had a place to go. It's not clear as to where this force is going.

To relate the above to the topic of today, we see a connection between systems engineering and the first-named four forces. That is, systems engineering needs to respond to these forces and be ahead of the power curve. Will it do so? That is not known, but this book offers some suggestions as to where the field might go.

STRESSES FOR SYSTEMS ENGINEERING AND INCOSE

Not too long ago, an ex-NASA administrator published an article entitled "How Do We Fix Systems Engineering" [2]. Griffin more than implies that systems engineering needs fixing; he gives examples of what has gone wrong, despite good intentions and sufficient spending. Here is a one liner that expresses Griffin's position:

> How is it that we continue to encounter failure of important and complex systems where everything thought to be necessary in the way of process control was done, and yet despite these efforts the systems failed?

And Griffin calls out the names, where it was presumed that systems engineering should have prevented a failure – Three Mile Island (TMI), Challenger, Columbia, power blackouts, and oil spills in the Gulf of Mexico. So what are the messages and potential solutions that Griffin has to offer?

Griffin suggests, first and foremost, that attention needs to be paid to creating the *elegant design*. He does not leave us wondering what that might be. Such an approach is characterized by the following features:

1. The design must work
2. It must be robust
3. It must be efficient
4. It must have a minimum of negative side effects

This author shall return to the topic of system design, in several later chapters. Meanwhile, Griffin goes on to deal with two other main topics. The first is apparent support of his argument from Robert Frosch [3], a previous NASA administrator, and how to achieve this notion of an elegant design.

The Frosch input [3] stresses that we must get way from peripheral tools and remember that "someone must be in control and must exercise personal management, knowledge and understanding", all with a sense of the art and excitement of true systems engineering.

On the matter of how to achieve elegant designs, Griffin points first to the academic community and the key role that they must play. Next, he suggests that the various communities of practice out there in industry move forward with action. He goes on to suggest that a third responsible area has to do with the structure and operation of the design team itself. The fourth and final area with respect to achieving elegant designs are our research establishments and paradigms. This author fundamentally agrees with this Griffin approach and has much more to say about design and its role in systems engineering.

INCOSE

Another stress is occurring within the body of International Council on Systems Engineering (INCOSE), responsible for much of what is happening in the overall field of systems engineering. To point to one source of stress, we look at an input from an INCOSE Fellow [4]. This Fellow claims that what INCOSE supports is "systems engineering management", and not "systems engineering". As such, it has little to offer as to hard-core engineering. He supports his thesis by pointing to such items as the V-Model, model-based systems engineering (MBSE), the dynamic behavior of elements, unresolved competency issues, and the systems approach. As of this writing, it is not clear as to how many people would agree with this Fellow's criticism, and how much activity it will engender. But it represents a stress that should be dealt with, one way or another, by INCOSE. This author suggests that there is considerable merit to the criticism and that appropriate responses by INCOSE are called for – perhaps even a committee to explore the matter more deeply and suggest changes that might be made.

The position of INCOSE with respect to the overall field of systems engineering can be represented by three main sources.

a. The journal from INCOSE, by name "systems engineering"
b. The systems engineering handbook, and
c. The INCOSE look at systems engineering by the year 2035

The Journal: Systems Engineering

This journal is an excellent purveyor of systems engineering knowledge, mostly as represented by INCOSE members. It is classical in the sense that it operates as a true journal, keeping away from cliques and advertisements of various types. The treatments are very much worth reading and in fact deal with the results of research in systems engineering.

The Systems Engineering Handbook [5]

This handbook is excellent in its scope and completeness. It certainly represents what INCOSE would call systems engineering. Highlights from the handbook include:

a. An overview of systems engineering
b. A definition of generic life cycle stages
c. Technical processes
d. Technical management processes
e. Agreement processes
f. Organizational project-enabling processes

Note the emphasis on process in this handbook, and also the fact that Griffin, the ex-administrator of NASA, makes the following observation:

> in a word, systems engineering is not fundamentally about "process", except in cases where such process is clearly lacking [2].

Systems Engineering Vision – 2035

INCOSE has documented its vision for the year 2035 [6]. Part of that documentation was a top-level statement of goals, namely:

a. To align systems engineering initiatives
b. To address future challenges
c. To broaden the base of systems engineering practitioners
d. To promote systems engineering research

As to where systems engineering needs to go, we provide here a one paragraph quote from INCOSE as to its necessary evolution:

> the practice of systems engineering must further evolve to support the demands of ever-increasing system complexity and enterprise competitiveness. Systems engineering will leverage the digital transformation in its tools and methods, and become largely model-based using digital representations of the system, simulation, and immersive technologies to rapidly adapt to change, and provide shared understanding of the system across its life cycle. The systems engineering practices will be based upon firm theoretical foundations. The systems engineering workforce will further expand the support the growing of model-based systems engineering needs from small, medium and large enterprises that span the broad range of industry applications.

INCOSE has also considered the nature of the focus to evolve the discipline. Three main topics have been considered in this respect, namely:

1. In order to evolve practices, one needs to explore:
   - The faster pace of change
   - Increasing complexity of systems
   - Affordable solutions
   - Agile, adaptable, and resilient systems
   - The challenges of tomorrow
2. Moving systems engineering to a cohesive discipline
3. There is a need to place emphasis on transforming systems engineering practices, such as:
   - MBSE, and its system models
   - Systems of systems
   - Complex systems

- Agile systems engineering
- Product line engineering
- Composable architectures
- Resilient and adaptable systems

NATIONAL DEFENSE INDUSTRIAL ASSOCIATION (NDIA) [7]

The Systems Engineering Division of the NDIA "advocates the widespread use of systems engineering in the DoD acquisition process in order to achieve affordable, supportable, and interoperable weapon systems that meet the needs of warfighters and provide the U. S. a technological advantage". Emphasis appears to be placed on a series of objectives that can be re-phased as follows:

1. Advancing the technical and business practices in systems engineering
2. Promoting excellence during system life cycles and across all disciplines
3. Transforming and modernizing Systems Engineer (SE) practices, while maintaining overall principles
4. Improving processes and practices to achieve required levels of performance
5. Pushing the boundaries of SE for system development and efficient life cycle management
6. Providing an industry perspective with respect to SE to government partners
7. Advocating for SE policy and guidance improvements

The NDIA takes it for granted that systems engineering is a crucial part of developing systems, when done properly. But they note, as above, that there is always a need for improvements.

THE OFFICE OF THE SECRETARY OF DEFENSE (OSD)

Major Austin Page (Air Force) explores the presumed need to "maximize the DoD's investment in systems engineering", starting from the Bell Labs' approach in the early 1940s [8]. This investment is largely dependent upon new technologies that allow various platforms to collect, process, and disseminate real-time information so that decision-makers can proceed with battle plans. This interoperability need requires a disciplined approach to both project management and systems engineering. Major Page claims that the effective use of systems engineering further requires learning from past experiences which includes both successes and failures. In the latter category, he looks at the so-called Nunn-McCurdy breaches that trigger reports when a program's unit cost exceeds certain thresholds. Major Page cites two specific program breaches:

1. The F-35 Lightning II Program in 2010
2. The Global Positioning System Next-Generation Operational Control System in 2016

He also cites the number of breaches from calendar years 2007 to 2015, as follows:

CALENDAR YEARS	CRITICAL BREACH	SIGNIFICANT BREACH
2007	1	4
2008	3	1
2009	7	1
2010	4	4
2011	4	–
2012	1	–
2013	2	2
2014	1	1
2015	1	1

Major Page points out that the Office of Defense, Assistant Secretary of Defense (SE) was chartered in 2011 as "the point of contact for policy, practice, and procedural matters relating to DoD Systems engineering" and "its key elements including technical risk management, software engineering, manufacturing and production, quality, standardization, and related disciplines". He also notes that the Systems Engineering Plan (SEP) has a template that addresses:

a. System architecture and interface control
b. Risk and opportunity management
c. Technical schedule and schedule risk assessment
d. Technical performance metrics and key performance indicators
e. Stakeholder management
f. Configuration and change management

    g. Technical reviews and their associated entrance and exit criteria
    h. Engineering tools, and
    i. Many other topics

A DoD systems engineering update was held within the OSD [9] that was quite specific about deliverables planned for FY 18. Significant items identified were as follows:

    a. Digital engineering
    b. The Modular Open Systems Approach
    c. Software
    d. Mission Engineering/System of Systems Engineering
    e. Sustainment engineering
    f. Trusted hardware and software assurance
    g. Cyber resilient weapon systems, and
    h. Modeling and simulation

Moving into the Biden administration, we find an initial but distinct emphasis on boosting innovation across the board within the DoD [10]. In order to achieve higher levels of innovation, a judgment was made that it is necessary to change the culture within the Pentagon. Emphasis was placed upon small business innovative research grants and modifying the day-to-day activities within the Pentagon.

THE SERC [11]

The SERC is a systems engineering research center, funded by the Office of the Secretary of Defense, Systems Engineering. The SERC has identified four primary research objectives, namely:

    1. Knowledge, data, and machine learning
    2. MBSE, decision-making, and integration
    3. Management, culture, and agility
    4. Security, trust, risk, and testing

Other areas specifically identified as a necessary part of the SE research agenda include [11]:

    1. Multiple topics relating to mission engineering
    2. Multiple topics relating to agile development and engineering at the scale of the enterprise

3. MBSE
4. Modeling and assessing system trust, security, and resilience
5. Development of pragmatic methods to deal with and to leverage increasing complexity and analytics, and leveraging machine learning
6. Digital engineering
7. Decision Support Frameworks

As with the other research activities, the presumption is that we will have made significant progress in the cited areas by a couple of decades down the road from the publishing of this book.

LOUGHBOROUGH UNIVERSITY [12]

The Loughborough University in the UK has been studying research challenges for systems engineering. They have concluded that there are five challenges as follows:

1. Business case, value proposition, and competency
2. Through life information and knowledge management
3. Modeling and simulation and total system representation
4. Systems engineering development environment and tools
5. System verification, validation, and assurance of extremely complex systems

The claim is, with respect to the above research areas, that they apply to both heterogeneous as well as autonomous systems.

THE DEFENSE INFORMATION SYSTEMS AGENCY (DISA) [13]

DISA focuses on all manner of information systems for the country at large. It has addressed its long-term approach in its strategic plan [13] that represents its overall mission as:

conducting DoD information network operations for the joint warfighter.

Under the category of operating and defending, it deals explicitly with:

a. Modernizing the infrastructure
b. End-user support
c. Computing
d. Defensive cyber operations
e. Readiness

DISA has demonstrated its interest in systems engineering by setting up and funding contracts in what they called SETI – systems engineering, technology, and innovation. This flexible contract vehicle gives the agency support in complex IT mission areas, incorporating innovation "into the requirements development and evaluation processes" [13].

SYSTEMS ENGINEERING BODY OF KNOWLEDGE

The Systems Engineering Body of Knowledge [14] gives us a glimpse into thinking about the field as we move on down the road. This includes the notion that:

a. Systems engineering will have a more encompassing foundation of theory and model-based methods, and
b. A core body of foundations will be defined and taught across academia.

Three system contexts are set forth, in particular:

1. A product context
2. A service context, and
3. An enterprise context.

The overall field will have to consider the role of autonomous agents as well as movement from a total control model to a shared responsibility model. A more organic model will involve growth, self-learning, self-organizing, and self-adapting. The systems engineer, in order to deal with these changes, will have to be more flexible and have the ability to adapt to new and shifting environments.

SPECIAL CITATIONS ACROSS THE BOARD

If we look across the board at the topics cited above, we find emphasis on the following areas, in terms of possible improvements that need to be made to make systems engineering more useful and effective (our "top baker's dozen" suggestions). Other suggestions are made in Chapter 10.

a. Knowledge, data, and information management
b. Interactive MBSE, including modeling and simulation
c. Agility, resilience, and adaptability in systems engineering procedures, processes, and practices
d. Establishing systems engineering as its own rigorous discipline
e. Security, trust, and risk analysis and mitigation
f. Affordability
g. SE development environments and tools
h. Handling the increased complexity of systems and systems of systems
i. More elegant design and architectures
j. Information Technology system readiness
k. Integrated modular approaches
l. Software systems engineering
m. Digital engineering

OVERVIEW OF CHAPTERS

Chapter 2. Tomorrow's Systems Engineer
Chapter 3. Kinds and Types Beyond the Tipping Point
Chapter 4. The Systems Approach and Thinking
Chapter 5. Systems Engineer as Executive VP
Chapter 6. Architecting and Modeling
Chapter 7. Software Re-Use and Agility
Chapter 8. Data and Decisions
Chapter 9. Miscellany
Chapter 10. Summary

REFERENCES

1. Eisner, H., "Reengineering Yourself and Your Company", Artech House, 2000.
2. Griffin, M. D., "How Do We Fix Systems Engineering?", Sixty-First Astronautical Conference, Prague, September–October 2010.
3. Frosch, R., "A Classic Look at Systems Engineering", IEEE International Conference, New York, March 26, 1969.
4. Wasson, C., "The Systems Engineering Conundrum: Where is the Engineering?", Paper # 94, INCOSE IS 2021, Session 4.3
5. Walden, et al., "Systems Engineering Handbook, Edition 4.
6. A World in Motion, Systems Engineering Vision – 2035, INCOSE.
7. Top Systems Engineering Issues in US Defense Industry, NDIA, Systems Engineering Division, 2016.
8. Page, A., "The Evolution of Systems Engineering in the US Department of Defense".
9. Baldwin, K., "DoD Systems Engineering Update", NDIA Systems Engineering Division Meeting, December 6, 2017.
10. Maucione, S., "How the Biden Administration can make DoD an Innovation Powerhouse", Federal News Network, January 2021. See https://federal news network.com
11. Primary Research Objectives, Systems Engineering Research Center (SERC), Report No. SERC-2018-TR-102, Stevens Institute of Technology, Hoboken, New Jersey.
12. Research Grand Challenges for Systems Engineering, Loughborough University, United Kingdom.
13. Strategic Plan 19–22, Defense Information Systems Agency (DISA).
14. Systems Engineering Body of Knowledge (SEBoK), Stevens Institute of Technology, SERC, Hoboken, New Jersey.

Tomorrow's Systems Engineer

2

Today's systems engineer remains strong with a systems perspective and support from INCOSE as well as from the DoD, Office of the Secretary. INCOSE, in particular, has put enormous energy into the growth and well-being of its members. So this author sees tomorrow's systems engineer as a kind of "super" systems engineer, with capabilities that are stronger than those of today. Here are some of the key aspects of what the super systems engineer can aspire to, as represented in the list below.

1. "Super" Systems Architect
2. Superior Team Leader on Systems Engineering Projects
3. Major Contributor to the Literature
4. One Step Down from Corporate Vice President
5. Entrepreneur

Super Systems Architect

It is essential that tomorrow's systems engineer become even stronger as the key architect of a variety of systems. The key word here is "synthesizer", and architecting remains the primary task of systems engineering. Examples of the super systems engineer that we should emulate are:

a. E. Rechtin
b. N. Augustine
c. B. Boehm
d. S. Ramo
e. A. Grove
f. T. Edison
g. R. Oppenheimer
h. Admiral H. Rickover

DOI: 10.1201/9781003269274-2

As they say, not a shabby collection of individuals. Some of the characteristics of these people are cited at the end of this chapter.

Superior Team Leader

One important role for tomorrow's systems engineer is building and running a project team. In this role the systems engineer gets to demonstrate his or her skills as a leader and also to make an immediate contribution to the well-being and capability of the enterprise. "Systems Engineer as Leader" may be a stretch in the minds of some, but our systems engineering community will be producing a large number of engineers who are capable of strong leadership at local levels, national levels, and even at international levels.

Contributor to Literature

We must insist that the systems engineer be able to contribute directly to the literature of systems engineering. This can be in the theory or practice of the field, and it may be in the form of papers as well as books.

One Step Down From Corporate Vice President

So where does the super systems engineer fit in the corporate structure? The answer here is just below his or her vice president. The answer argues that this systems engineer is more likely to rise even above that. The role, or job, of this systems engineer, however, is to support the agenda of his or her vice president and the corporation.

Entrepreneur

This engineer has been given the charter to help his or her vice president in expanding the current line of business (LOB), not start another LOB.

What Are the Success Factors for the Systems Engineer?

Some while ago, the author formulated a list of features of the successful systems engineer [1]. A very brief citation of these factors is provided below:

> *Synthesizer:* This attribute remains critical in terms of what the super systems engineer is expected to do. It is also at the heart of systems architecting.

Listener: The best way of gathering information, aside from reading reports, is to be listening, all the time. When the mouth is moving, the chances are no new information is being absorbed.

Curious/Systems Thinker: The key capability for this engineer is and remains systems thinking. This is further explored in Chapter 6. Also, for this systems engineer, curiosity is a strong driver.

Manager/Leader: This systems engineer knows when and how to move from a purely technical role to a management role. The stronger of these engineers will exhibit the attributes of a leader.

Expert/ESEP: This expertise pertains to the well-defined features of the nut and bolts of systems engineering (as documented in Mil Std 15288). Awarding the ESEP level of capability is verification by INCOSE.

Expert/Domain Knowledge: Expertise, for the successful systems engineer is expected to be confirmed for at least one outside domain. Some examples of the latter include:

a. Communications
b. Missile defense
c. Transportation
d. Information theory
e. Cybersecurity

Perseverer: Staying with the task is recognized as an important ingredient in success. Edison lived and breathed this attribute. High achievers demonstrate it everyday.

In the process of identifying success factors, INCOSE fellows were queried for their inputs [1]. Here is a sampling of such inputs that appeared to be especially relevant:

a. Interdisciplinary growth and understanding
b. Engages in computer-based systems engineering
c. Creative architectural design
d. Team leadership skills
e. Negotiation skills, with customer and management of your organization
f. Able to address and solve risks inherent in building a large system
g. Understanding the whole system and seeing the big picture
h. Problem domain knowledge
i. A holistic perspective
j. Embracing continuous learning
k. Tolerance for ambiguity
l. Technical leadership

Yet another set of inputs was obtained, in this case in the form of three articles from the literature [1]. These are re-examined below.

NEW ENGINEER ARTICLE

This citation begins with "a team player", suggesting that the successful systems engineer must know how to participate in a team to the advantage of everyone. Second, this engineer embraces the "Fifth Disciple" approach set forth by Senge. This engineer is creative, thinking outside the box when and if necessary. This engineer is capably of "problem-solving", known to be a most valued attribute by many employees. This engineer has a strong analytic ability – highly valued in many aspects of building new systems. As implied by some of the above, this engineer has strong skills in communication. With whom? With everyone. This engineer is a logical thinker, related sideways to "systems thinking". This engineer pays attention to detail. This book's author would not necessarily support this choice since a part of the systems approach is to think upward and sideways, not leaving a lot of time and energy for detail which might be construed as thinking downward. This engineer enjoys mathematics and is able to use it with great facility. This engineer shows definite signs of leadership, going beyond the citation as a team player.

SPEC INNOVATIONS ARTICLE

This engineer is comfortable with being patient and persevering. In these terms, they go together or at least are highly correlated. This engineer knows when he or she is done. This may well connect to "the perfect is the enemy of sufficiently good". This enemy, if highly analytical, executes things with ease and high productivity. This engineer knows about and can use systems engineering software tools. This engineer has strong organizational skills, which help in most management tasks. "Ability to see the small picture" is an attribute of this systems engineer. Here again, this book's author does not necessarily agree with this choice. This engineer also sees the big picture (that's more like it). This engineer has a well-rounded background. This engineer has strong communications skills, necessary as a team player and project manager. This engineer has the ability to lead, follow, and work well in a team. It's teams that build systems, and this skill solidifies the effectiveness of the team.

NEWS PATROLLING ARTICLE

This systems engineer has a strong technical background. This, of course, includes various types of engineering as well as the sciences. This engineer can handle multiple responsibilities. A typical example is to be the team leader on one project and a team player on another. This engineer has an analytical mind, with considerable training in analysis. This understanding knows something about the Law. Short of being a lawyer, this engineer gravitates to problems and systems pertaining to the Law, and does well with his or her baseline understanding. Another dimension is the set of terms and conditions that relate so the system at hand. This engineer is able to make quality decisions. These decisions could be at the project leader level or at the level of a design engineer. This engineer is able to communicate both down and sideways. This engineer is humble with respect to his or her contributions. This systems engineer will need to be self-motivated. Doing important and interesting work should be motivation enough. This engineer is able to diagnose problems. An important skill since all projects, even small ones, have an array of problems that need to be addressed. Is good at listening and implementing ideas that go beyond his or hers.

MANAGEMENT SKILLS OF THE SUPER SYSTEMS ENGINEER

Four purely management skills come to mind, namely:

1. Building a high performance integrated team
2. Assuring that the system is built fast and more cheaply
3. Interfacing with stakeholders
4. Interfacing with top management and Congress (if appropriate)

In the first role, we envision the systems engineer as the project manager of an important project, whether from industry or government. The project team may likely need to be assembled from scratch. Starting with a clean sheet of paper, the systems engineer will likely need to do a lot of interviewing from within and without. Getting members to agree to be part of the team is only the beginning. Training sessions follow immediately, and a project plan would have to be assembled and agreed upon.

The attributes of an integrated high-performance team are not new. The military has taken a strong lead in developing high-performance teams as demonstrated by this enumeration of the skills of such team members [2]:

a. Establish strong competence
b. Develop trust relationships
c. Formulate common understandings
d. Ensure commander's intent
e. Issue mission orders
f. Assure disciplined approach
g. Consider balanced risk
h. Establish effective communications

In the second role, we can insist that the superior systems engineer acting as a project manager be able to bring the system home ahead of schedule and at or reduced costs. Both are contingencies that reduce risk and the systems engineer has that awareness from the beginning. It's not just getting the job done; it's doing so in a superior manner.

In the third role, the systems engineer must interact with any and all stakeholders when dealing with an important system. That implies superior communication skills, demonstrated by timely briefings that provide key information and transparency

The fourth management ole is another level and focus in terms of interactions and interfaces with top management and Congress. If in the military that includes a host of generals, and if in the civil area, it includes members of Congress.

The three technical roles that need to be executed by the superior systems engineer are:

1. Developing and assuring a superior system architecture
2. Achieving high technical performance from the system itself
3. Contributing to the field of systems engineering

In the first technical role, nothing is more important that the correct technical system architecture. A robust and well thought out architecture is the key to a long-terms success.

In the second technical role, all levels of technical performance comes immediately after the correct technical architecture. It's the "better" part of

Faster and Cheaper, but from a technical point of view. It's the system doing its thing in terms of satisfying all technical specifications and requirements for the system.

The third technical role is to insist upon making one or more technical contributions to the overall field of systems engineering. Such contributions would have special meaning when set forth from an important system development.

The top-level systems engineer responsible for an important system may be illustrated by the case of Robert Oppenheimer. He would be a prototypical example of what we would be striving for. See the thumbnail sketch later in this chapter.

The role and capability of the systems engineer as problem solver may be examined in the context of his or her ability, in general, to address and solve problems. Here, we turn to a previous work by the author, in the domain of general problem-solving [3]. In that connection we recall the dozen primary kinds of problems to be addressed that would assure a high level of problem-solving expertise, namely:

1. N-Step Problem-Solving
2. Reductionism
3. Modeling and Simulation
4. Lateral Thinking
5. Total Systems Intervention
6. Generalized Systems Thinking
7. Design (IDEO) Approach
8. Expert Systems
9. Mathematics and Statistics
10. DoD – Suggested
11. Decision Support Systems
12. Cost Effectiveness Analyses

What this author discovered in writing his book on problem-solving is that it's mostly (but not always) a two-step process. Step one is you look at the problem to be addressed and figure out what type or kind of problem it is. After you've classified the problem, you let the problem type or kind assist you with a solution, trying hard to not invent the wheel or solve a problem that's already been solved. So, for example, if you've got a problem in optimization, you look for existing optimization software that is likely to be applicable to your problem. Or if you have a process flow problem you might go to the Modeling and Simulation software and instantiate your problem in that methodology. The two-step notion helps to get you to an answer relatively quickly and, as suggested, without re-inventing the wheel. You don't start from zero, but from a place in which a lot of progress has already been made.

As this author went back and skimmed the list of approaches, he found that, as a practical matter, there were about 50 kinds or types of problems represented. The remainder of this chapter deals with a very brief exploration of these 50 kinds of problems, followed by commentary regarding the connections to systems engineering.

13. By fable
14. By algorithm
15. By the eight disciplines
16. By analogy
17. By deductive logic
18. By appreciative inquiry
19. Using synectics
20. By grounded theory
21. Using morphological analysis
22. Using TRIZ
23. Using abstraction
24. Using brainstorming
25. Using means-ends analysis
26. Using system dynamics
27. Using collective insight
28. Using visualization
29. Using divide and conquer techniques
30. Using 2×2 matrix analysis
31. Using re-engineering procedures
32. Using reduced clock speed
33. Using first principles
34. Using value focused
35. Using mindset
36. Using debating techniques
37. Using Osborn-Parnes process
38. Using disruptive thinking
39. Using critical thinking
40. Using systems thinking
41. Using river crossing methods
42. Using coin weighing procedures
43. Using risk analysis
44. Using message coding
45. Using parameter dependency diagramming
46. Using transformational calculus
47. Using Leontief logic
48. Using focus-diffuse methods

49. By institutional methods
50. Based upon acronyms

A few words on the "top dozen" kinds of problem-solving approach follow.

By N-Step Problem-Solving

The enumeration of specific steps was started with the Ford Motor Company, and later picked up by other companies as well as the military. An example is the 5 step process articulated by this author as a generic process, listed below:

1. State the problem
2. Identify key variables and dependencies
3. Identify alternative solutions/approaches
4. Select and implement best alternative
5. Document and present results

By Reductionism

This approach calls for breaking the problem into pieces, solving each piece, and then putting the solutions back together. Many such approaches are based upon conditional probability breakdowns.

By Modeling and Simulation (M & S)

Often applied to complex processes and related to the time lines of systems. Simulation is a special case of model that can be time driven or event driven.

By Lateral Thinking

Looking "sideways" for a solution. The metaphor is digging holes and looking sideways at a hole not yet explored. That new hole is lateral thinking.

By Total Systems Intervention (TSI)

This method was devised by two researchers from the UK, with the names Flood and Jackson. The former has described TSI as "an approach to problem-solving in any organization that stands firm with the original holistic intent of systems thinking".

By Generalized Systems Thinking

This approach is typically based upon thinking "upward", and recognizing that the problem you have is part of a larger "systems" problem.

By the Design Approach

A method of IDEO (company) and the author of a book by Tim Brown titled Change by Design. Leans heavily upon building prototypes and team approaches.

By Expert Systems

This approach requires the availability of an expert system, and its key ingredients, a data repository and an inference engine. The expert system is typically instantiated in software.

By Mathematics and Statistics

This method looks to the fundamental fields of math and statistics to find a solution in their varied aspects.

By DoD Suggested

Here we lean on the many years of problem-solving, especially by the military. An example is the use of surveillance theory in air defense systems.

By Decision Support Systems

This method requires a team approach as well as software that has one or more built-in algorithms that support that team.

By Cost-Effectiveness Analysis

This specific method evaluates costs and measures of effectiveness for well-defined alternatives. These are bases for finding what is called a cost-effective solution.

So we note that a "base" of about 50 kinds of approaches gives one a sense of the potential effectiveness of the overall approach.

EXAMPLES OF HIGHLY SUCCESSFUL SYSTEMS ENGINEERS

E. Rechtin

Rechtin was a triple threat as a systems engineer: he was president of the Aerospace Corporation, headed DARPA, and served as professor at University of Southern California (USC). He made seminal contributions with his book on systems architecting and heuristics pertaining to building systems. He also applied his concepts to the overall organization of enterprises.

N. Augustine

Norman Augustine might be considered a quadruple threat – (1) President of Lockheed Martin, (2) Secretary of the Army, (3) professor at Princeton, and (4) super consultant for government and industry. His book "Augustine's Laws" was a best seller and contained much wisdom. He followed that treatise with "Augustine's Travels".

B. Boehm

Barry Boehm has been our leading software engineer for years, contributing to the issue of software estimation. A mathematician by training, he has led the charge toward solving one of our most serious problems, that is, building trusted software for our large-scale systems. He serves as deputy for the SERC contract held by the Stevens Institute of Technology.

S. Ramo

Simon Ramo was this country's leader in missiles and space systems. He is the "R" in TRW, and advised the government on important defense issues for several decades. He was the author of many books pertaining to engineering and management. He received many awards including the Medal of Freedom from President Reagan.

A. Grove

Andy Grove was a key executive at Intel and, as such, kept the chips flowing and of high quality. He's a good example of how "immigrants" are able to find their way and make serious contributions to our technology and economy. He funded the Grove School of Engineering at the City College of New York where he obtained his Bachelor's degree.

T. A. Edison

Thomas A. Edison, the inventor from Menlo Park, New Jersey, had over 1000 patents to his name. It was his fine touch and inventiveness that led to the phonograph, light bulb, motion pictures, and large-scale power generation and delivery systems. The Edison power company is a highlight across the New York City skyline. He is reported to be the originator of the phrase "invention is one percent inspiration and 99 percent perspiration".

R. Oppenheimer

Oppenheimer was given the assignment of managing the Manhattan project which was the task of the century. Its purpose was to develop the first atomic bomb, while managing a group of the smartest physicists in the country. He was given lots of freedom to do as he saw fit by General Leslie Groves, who knew how to deal with Oppenheimer, himself a very intelligent and independent physicist. Oppenheimer became known as the "father of the atomic bomb" and also a proponent of the notion "now I am become death, the destroyer of worlds".

Admiral H. Rickover

Admiral Rickover was a key player in the Navy, building the nuclear Navy and the missile carrying nuclear submarines. Nominally, Adm. Rickover reported to the Navy head of the Bureau of Ships, but had enormous influence with the Chiefs of Staff as well as important members of Congress.

He had his own unique way of testing recruits for the nuclear submarines, and many tell the story of how he challenged them during the interviewing process.

REFERENCES

1. Eisner, H., "What Makes a Systems Engineer Successful?", CRC Press, 2021.
2. Mission Command, see https://www.co.kearney.com
3. Eisner, H., "Problem Solving – Leaning on new Thinking Skills", CRC Press, 2021.

The Systems Engineer as Executive VP

3

Tomorrow's systems engineer must be excellent in terms of management as well as technical skills. Further, he or she must be of executive quality and capability, as in the executive Vice President (VP) of any enterprise. This capability is beyond just being a good manager; it occupies one of the top corporate executive positions.

The top-level systems engineer responsible for an important system may be illustrated by the case of Robert Oppenheimer. He would be a prototypical example of what we would be striving for. Here are some quotes from his biography that further describe the person and the capability [1]:

> Teller: He showed a refined, sure, informal touch
>
> Bethe: He never dictated what should be done. He brought out the best in all of us ... His grasp of problems was immediate
>
> It helped that Oppenheimer could turn on – and turn off – his personal charm

There is little doubt that Oppenheimer could serve as executive VP of just about any organization, the more technical, the better. He led by example, by his force of personality. He engendered a sense of trust and of motivation for the rest of the company or team.

THE PROJECT TRIUMVIRATE

Another "model" of the role of tomorrow's systems engineer can be seen in terms of the three person management at the project level [2]. This has been called the triumvirate and consists of the Project Leader, Chief Engineer, and

DOI: 10.1201/9781003269274-3

Project Controller. Our systems engineer easily fits into the chief engineer position, and should have no problem with the responsibilities of that position, which have been identified as:

a. Establishing the overall technical approach
b. Evaluating alternative architectures
c. Developing a preferred architecture
d. Implementing a repeatable systems engineering process
e. Implementing a repeatable software engineering process
f. Oversee the use of computer tools and aids
g. Serving as technical coach and team builder
h. Carrying out technical review sessions
i. Attempting to minimize overall schedule
j. Ultimately building a cost-effective system that satisfies the stated requirements

Each of the above deserves a brief comment.

Architectures

Formulating an architecture for the system in question is a critical element for the chief systems engineer.

Systems Engineering Process

We have reached a point at which the systems engineering process is well known and well documented. The chief systems engineer's job, in part, is to assure that such a process is faithfully executed.

Computer Tools and Aids

These need to be brought to bear by selected members of the systems engineering teams. That's what Model-Based Systems Engineering is all about.

Technical Coach and Team Builder

The chief systems engineer is able to coach individuals and, in addition, build a highly functional team.

Technical Review Sessions

The chief systems engineer calls for and chairs these sessions.

Schedules

This important indicator of progress needs constant attention.

Cost-Effective Solution

The overall system must represent a "cost-effective" solution in terms of satisfying the system requirements.

CONNECTIONS TO THE SYSTEMS ENGINEER OF TOMORROW

Tomorrow's systems engineer will be capable of being an excellent problem solver once he or she has received sufficient training in this area. It takes practice, as they say, and as implied by those who have studied this matter. The bottom line is that the systems engineer some 20 years from now will have the background and capability to serve as the executive VP of any enterprise, including those with high technical content. This would place the systems engineer just behind the Chief Executive Officer (CEO) or president as responsible for all that matters in the enterprise. This is certainly an expanded role as compared with that of today. Perhaps the relevant question then becomes – will today's system engineer be able to transition to the Executive of tomorrow?

OTHER EXAMPLES OF SYSTEMS ENGINEER AS EXECUTIVE

Earlier in this chapter we cited Robert Oppenheimer as an example of the "super" systems engineer for tomorrow. Here are four more examples of this type of systems engineer.

Norman Augustine

Mr. Augustine was (and remains) a quadruple threat: superior aeronautical engineer, President of Lockheed-Martin (no small job), professor and trustee at Princeton University, advisor to government on important defense and aeronautics issues, author, receiver of some 14 honorary doctorate degrees, receiver of special awards such as the National Medal of Technology, von Karman Wings Award, a member of the national Academy of Arts and Sciences, and Secretary of the Army.

Eberhardt Rechtin

Dr. Rechtin also was a quadruple threat: Superior communications engineer, writer of seminal text on systems architecting which contained important heuristics for the systems engineer, head of Director of Defense Research and Engineering (DDR&E) in government, professor at USC in California, and President of the Aerospace Corporation.

Simon Ramo

The "R" in TRW, another quadruple threat, superior communications engineer; received many awards, including the Presidential Medal of Freedom, the Hill Lifetime Space Achievement Award, the IEEE Founders Meal, and the John Fritz Medal; left Hughes to form the Ramo-Wooldridge Corp, which later became TRW; developed the Falcon missile; worked with the Air Force to deliver integrated Radar and aircraft fire-control systems; served as Chairman of the President's Advisory Committee on Science and Technology.

Admiral Hyman Rickover

Credited with building the nuclear Navy, especially the Nuclear-capable submarine, known as the "Father of the Nuclear Navy", Served as flag rank of the Navy for almost 30 years; served as longest -serving naval officer; received Presidential Medal of Freedom, Legion of Merit, Congressional Gold Medal, Navy Distinguished Service Medal, and the Enrico Fermi Award.

Irwin Jacobs

Dr. Jacobs was a co-founder of Qualcomm where he made his considerable fortune. He also taught engineering at both Massachusetts Institute of Technology (MIT) and USC, having written an important book entitled "Principles of Communications Engineering". He received the Marconi Prize, the IEEE Medal of Honor, and the IEEE Alexander Graham Bell Medal, among others. He served as chairman of the Salk Institute. He was elected as a member of the National Academy of Engineering in 1982 for contributions to communications theory and practice and is on the international advisory board for the Israel Institute of Technology.

Andrew Viterbi

Co-founded Qualcomm (with Irwin Jacobs); invented the Viterbi algorithm dealing with code-division multiple access systems; served on Board of trustees for USC; served as professor of engineering at both University of California, Los Angeles (UCLA) and USC and was elected to the National Academy of Engineering in 1978; received the Golden Jubilee Award for Technological Innovation from the IEEE; awarded the Benjamin Franklin Medal in Electrical Engineering; received the IEEE Medal of honor.

Andy Grove

Dr. Grove served as CEO of the Intel Corporation; and also as first Chief Operating Officer (COO) and third employee; named TIME Man of the Year (1997); wrote a seminal book on physics and the technology of semiconductor devices and one on High Output Management; founded the Intel Corporation with Robert Noyce and Gordon Moore (of noted Moore's Law); introduced the 32-bit microprocessor; CEO of the year for Chief Executive Magazine.

The above "larger than life" systems engineers represent the very best that the US has had to offer. Thus they are to be emulated in terms of technical and management abilities and dedication to their professions. One can imagine that some couple of decades down the road there will be several more "super" systems engineers that have achieved what the above-cited engineers have been able to achieve.

In addition to being considered Executive material as technical and management experts, these excellent engineers qualify in technical leadership roles, as defined by a noteworthy author [4]. She documented several attributes of effective leaders, under the categories of the following:

a. Traditional attitudes
b. Great man theories
c. Transactional and transformational leadership styles
d. The contingency model of leadership

 e. Situational theory
 f. The path-goal model
 g. Authentic leadership
 h. Complexity leadership
 i. Followership
 j. Competencies
 k. Behavioral attributes
 l. Leadership, communication, and management styles

REFERENCES

1. Bird, K. and Sherwin, M., "American Prometheus – The Triumph and Tragedy of J. Robert Oppenheimer", Vintage Books, 2006.
2. Eisner, H., "Problem Solving - Leaning on New Thinking Skills", CRC Press, 2021.
3. Eisner, H., "Essentials of Project and Systems Engineering Management", John Wiley, 2008.
4. Davidz, H., "Technical Roles in Systems Engineering", see also https://www.sebokwiki.org/wiki/Technical_Leadership_in_Systems_Engineering

Kinds and Types Beyond the Tipping Point

4

We start with the a priori premise that the systems engineer of tomorrow will need to develop skills and approaches that go beyond the systems engineer of today. And we hypothesize that the new skills and approach are likely to fall in the domain of "design". That is, the systems engineer of tomorrow needs to be a better designer than that of today.

Another word for design is "synthesize". This observer of the scene of systems engineering sees major curriculum strengths in analysis, but not synthesis. Today's systems engineer is excellent at analysis, but not synthesis. Accept that premise, if you will, before you move on.

SYNTHESIS

A leading researcher and technologist [1] has declared that synthesis will "supplement analysis as a major thrust of technological society". Synthesis involves putting pieces together according to some overall plan for functionality. Synthesis is more-or-less the same as "design". It is also a synonym for creating or constructing. In general, for this author, it has not been given enough attention in our learning institutions, with the possible exception of the profession of architect.

DOI: 10.1201/9781003269274-4

ANALYSIS

The subject of "analysis" is recognized as the opposite of "synthesis". In analysis, one usually starts with a given configuration, such as a circuit, or a detailed design, and is asked to compute the voltages, currents, power, and the like. In communications and transportation, the analyst is asked to compute values such as grade of service, waiting times, trip times, and other significant variables. The basic point in analysis is that the analyst starts with a given configuration and is asked to "solve for" these key variables. Where did the configurations come from? They came from the designers, the synthesizers. In some sense, the analysts are checking to see if the synthesizers have done it correctly – whether they have designed a system that satisfies the requirements on both qualitative and quantitative levels.

Two well-known treatments dealing with analysis are the books by Blanchard and Fabrycky [2], and Bill Sherer et al. [3]. The former is a classic, dealing with such topics as queuing theory, control systems, and reliability. The latter explains what is meant by the "ten golden rules of systems analysis" among other analysis topics.

REPLICATIVE DESIGN

We consider a new phrase, replicative design, and use it as a new skill and approach to system design. An example of such a design process can be illustrated with this statement:

> The systems shall be composed of 18 systems of kind A, 12 systems of type B, and 7 systems of essence C.

This suggests at least three overall kinds, types, and essences, which in turn suggests standard modules, and yes, that's where replicative design would be headed – to standardized modules that are easy to produce for tomorrow's systems.

In broad terms, we are looking to evolve standardized forms of design that have advantages in reliability, cost and producibility, and other "ilities". We are looking to improve the overall design process through deep standardization.

This will not mean new and better design skills and approaches are just around the corner. It is likely that this kind of new approach will take decades to develop.

One perspective regarding this approach is that we are looking to build many new systems not with the resources of many but with the resources of few. Not everything is a "one at a time unique project" but rather let's replicate what we've done before. That new satellite system is composed of several standard modules, all of which have been done before and continue to be part of this re-use concept.

We are seeking a new theory of "build once, use many times" for assurance of the design process and overall design superiority. Today's world is one that leads to increases in schedule, cost, and performance of systems, as a generality. Tomorrow's world holds out hope for Dan Goldin's suggestion of "faster, cheaper, better" [4]. Goldin was the NASA Administrator from April 1992 to November 2001 and founder of Cold Canyon AI, an innovation company. His vision of NASA and its missions was extensive, and recognized positively by the various space and aeronautics stakeholders across the country.

THEORIES OF DESIGN

There is a sparse theory of design that explores how to do design, and the types of thinking that goes into design. Herb Simon is a name [5] that one will find as associated with design. The literature shows several sources dealing with design, with a dominant position held by architects.

Design, these days, exists at the hands-on levels (such as mechanical system design) and also at the atomic level. In the latter case, we have moved into astonishing number of transistors on a chip [6]. According to the IEEE Spectrum magazine, we have reached the level of trillions of transistors on a chip. This has meant that the system and sub-system designers need to look more deeply into design methods and technologies.

A new design process that embodies the principle of "build once, use any times" is a second cousin to the DOTSS procedure. DOTSSS stands for Developer off the shelf systems and suggests that once a certified system is built one can use it many times [7]. One presentation of the features of DOTSS provided an overall improvement of 2500%. That's not 25%, but 100 times greater than that. More information regarding the DOTSS idea and approach is provided in Chapter 7.

The reference here to *kinds* of systems, *types* of systems, and *essences* of systems suggests the need on the part of the systems engineer to spend more time and energy in the study and application of taxonomy or classification. In other words, such investigation needs to be part and parcel of systems engineering which is not today's orientation or perspective. In tomorrow's world of systems engineering, taxonomy will play an increasingly important role, especially with respect to system design.

MODULAR OPEN SYSTEMS APPROACH (MOSA)

Related to the above is the so-called MOSA – the Modular Open Systems Approach. That has been receiving considerable attention in the past several years. All of that is appropriate and is discussed in greater detail in Chapter 6.

BEYOND THE TIPPING POINT

In this section we explore the vagaries of Malcolm Gladwell's tipping point [8] and other related perspectives.

The great success of Gladwell's specifics as well as his new ideas can be activities dealing most intelligently with:

a. The soundness of the idea itself
b. The articulation of the idea as well as its derivatives
c. His proof which lie in the many chapters of examples

The systems engineer needs to pay special attention to Mr. Gladwell and the many messages in his books. The reason, which may not be apparent, is that Mr. Gladwell is exploring systems behavior and that is precisely what the systems engineer needs to know. Mr. Gladwell gives us many explanations and examples of the tipping point; and the systems engineer is in a position to use this information as he is building systems. For example, the systems engineer is well familiar with saturation effects and when the "curve turns over". So the tipping point enters the arena of the familiar to the systems engineer.

Gladwell defines the tipping point as the moment of critical mass, the threshold and the boiling point. The systems engineer is likely to put forth a definition something like the point at which a system exhibits behavior in which the addition of new effort leads to a less than linear increase in consequence. Looking at a plot of the effort function we see a drop in the y axis as more and more effort is applied to the x-axis. The function "droops" as it departs from linearity.

Engineers are generally familiar with saturation effects, as for example with the avalanche diode. This diode breaks down when a specific reverse bias voltage is applied. It is used to protect circuits and also serve as a noise generator when such is necessary.

Moving beyond the tipping point there are numerous paths we can take. Three potential paths often appear.

a. Continued linear motion
b. Exponential upward motion
c. Exponential downward motion

In all cases the post-tipping point activity is not arbitrary. There are reasons for the (a), (b), and (c) alternatives and these reasons are largely explained by forces or lack thereof. So if the post tipping point moves exponentially up, it will be due to upward forces. In the case of exponentially downward, the forces will be down, in the aggregate. In the case of continued linear motion, the new forces will net to zero.

Activity beyond the tipping point, in all cases, is a reflection of system (or sub-system) behavior, and therefore, lies in the domain of the systems engineer. Indeed, one might say that all such activity and trajectory remain in the interest of the systems engineer. Why is the post-tipping point going up, down or sideways. All these are questions the systems engineer must continue to explore. This I why and how Gladwell's interest and that of the systems engineer coincide.

OTHER GLADWELL INTERESTS

A second treatise from Gladwell deals with the subject of "Blink" [9]. Gladwell puts forth the proposition that critical thoughts and actions often take place literally in the blink of an eye, and that such thoughts and actions are often extremely good and appropriate. An example of how this might be reflected in the interest of the systems engineer is the field of human factors. More specifically the systems engineer need this kind of perspective to design appropriate training programs for the astronaut and the pilot under combat stress. For the pilot especially, the blink of an eye might be enough to tell the tale between failure and success.

In the case of Gladwell's *Outliers* [10], at least two results need to be part of the systems engineers competency areas. The first is the apparent fact that it takes 10,000 hours of appropriate practice to achieve competency. Where might this apply? As above, possibly in the domain of the fighter aircraft pilot. Another area has to do with threshold and intelligence. Gladwell suggests that there is a threshold such that above the threshold will not help in terms of performance.

So what is it that might connect Gladwell with the systems engineer? This commentary suggests that there is a connection and it might well be immediate. As such we might well be ready to modify the systems engineering training along the lines of a requirement to stay in touch with the literature. An expansion of this idea is provided below.

THE LITERATURE CONNECTION

Keeping up with one's field, and related fields, should be mandatory in the INCOSE world, and the systems engineering domain generally.

To illustrate the concept, here is a list of a dozen publications that should be of special interest to the systems engineer:

1. INCOSE Handbook [11]
2. ISO excellence [12]
3. From Good to Great [13]
4. Oppenheimer Biography [14]
5. Bell Labs' Systems Engineering book [15]
6. The innovators, from Isaacson [16]
7. Duckworth's Grit [17]
8. Zen and the Art of Motorcycle Maintenance [18]
9. Dr. Deming and his Philosophy [19]
10. Eberhardt Rechtin and his Heuristics [20]
11. Reengineering the Corporation [21]
12. Built to Last [22]

These key subjects keep the systems engineer in touch with a broader range of topics that will likely influence his or her approach to the tasks at hand, over the years.

INCOSE Handbook

The fourth edition of the INCOSE handbook may not be wonderful bedtime reading, but it represents an up-to-date treatise on the fundamentals of systems engineering. It is inclusive, as it covers the subject top to bottom, and needs to be available to the systems engineer as he or she goes about on a systems project. Key areas covered include:

    a. An Overview of Systems Engineering
    b. Generic Life Cycle Phases
    c. Technical Processes
    d. Technical Management Processes
    e. Project Enabling Processes
    f. Cross-Cutting Methods
    g. Specialty Engineering

In Search of Excellence

The authors, Peters and Waterman, set out to discover the attributes of companies that were taken to be excellent. They wound up with quite a few attributes, as listed below:

    a. Able to manage in the face of ambiguity and paradox
    b. A bias for action (the first of eight basics that follow)
    c. Close to the customer
    d. Autonomy and entrepreneurship
    e. Productivity through people
    f. Hands-on and value driven
    g. Stick to the knitting
    h. Simple form and lean staff
    i. Simultaneous loose-tight properties

This author leaves it to the reader to elaborate on this ground-braking list.

From Good to Great

Collins makes the following main points: (1) going from good to great requires "Level 5" leaders, (2) need to embrace a set of practical disciplines, (3) use technology as an accelerator, (4) radical change will fail, so avoid, and (5) drive toward simplicity. Many companies have figured it out, and applied these points successfully, such as Fannie Mae, Gillette, Kimberly-Cark, Kroger, Pitney Bowes, Walgreens, and Wells Fargo.

Robert Oppenheimer

Oppenheimer served as a professor of physics at UC-Berkeley and was called upon to run the Manhattan project for our country. This was to develop the atomic bomb that was ultimately used to bomb Hiroshima and Nagasaki. This massive destruction led Oppenheimer to fight against the Hydrogen bomb and his citation as well as embracing of these words from the Bhagavad Gita: "Now I Am Become Death, the destroyer of worlds".

Bell Labs

Under the leadership of not-well-known Mervin Kelly, Bell Labs was the leading innovator in the country from the 30s to the 70s. The familiar names and contributors of Shockley, Shannon, and Pierce led the charge with the transistor, information theory, and satellite communications, respectively. Many colleagues of this author applied for a job at Bell Labs after graduation, with the desire to work with the best.

The Innovators

Walter Isaacson tackled several interesting topics, one of which was profiling "The Innovators". The list of same included von Neumann, Larry Page, Sergei Brin, Bill Gates, Marc Andreessen, Vint Cerf, Bob Kahn, Steve Jobs, Alan Kay, and IBM's Deep Blue and Watson computers. And beyond the names were the technologies they dreamed about and then developed.

Grit

Dr. Duckworth explains what grit is all about and why it's important. It's almost as if she were writing about perseverance and the Edison of Menlo Park. Some highlights – Achievement is made more possible when one takes into account these two formulae: talent × effort = skill and skill × effort = achievement (i.e., effort counts twice). How much effort is suggested? She points to repetitive high-quality practice of 10,000 times, just as the practice is productive and well founded.

Zen and the Art of Motorcycle Maintenance

The literature says that Robert Pirsig queried more than 100 publishers before he found one that said "yes". The result, of course, was his book about Zen, maintenance, and the entire subject of quality. The Zen book is sub-titled "an inquiry into values" and that's exactly what it is.

Out of the Crisis

Dr. Deming is well known for his 14 points that sum up his approach to achieving quality products and services. In addition, he set forth a "system of profound knowledge" that was a "holistic approach to leadership and management". It was also a theory of self-improvement in the area of total quality management.

Eberhardt Rechtin and His Heuristics

Dr. Rechtin made his way in the three sectors of industry, government, and academia, distinguishing himself in all. But in this connection, we point to his excellent list of heuristics in regard to building systems. A short sample of some of these is cited below:

a. Important software mistakes are made on the first day
b. For new systems, expect the unexpected
c. Choose among alternative architectures
d. No system can be optimal for all parties
e. Recognize Pareto's Law
f. KISS (keep it simple stupid)
g. Keep system requirements under challenge
h. Maintain options as long as possible

Reengineering the Corporation

Hammer and Champy wrote this blockbuster book about how to reengineer your corporation. Here are some of their themes:

a. There are four key words: fundamental, radical, dramatic, and processes
b. Process orientation; rethinking business processes
c. Checks and controls are reduced

d. Rule breaking is permitted

e. Creative use of information technology

f. Start with a clean sheet of paper

Built to Last

This treatise is sub-titled "Successful Habits of Visionary Companies". A role of the systems engineer down the road in years is likely to be to help to build an enterprise (or two). He or she will benefit from what it takes to be a "visionary company". The authors compared visionary companies against other companies. That led to a listing of the visionary companies and their key ideologies, as abstracted in Table 4.1.

TABLE 4.1 Visionary companies and selected ideologies

VISIONARY COMPANY	TWO SELECTED IDEOLOGIES FOR EACH COMPANY
1. American Express	a. Heroic customer service
	b. Worldwide reliability of services
2. Boeing	a. Leading edge of aeronautics
	b. Product safety and quality
3. Ford	a. Basic honesty and integrity
	b. People as the source of our strength
4. General Electric	a. Improved quality of life through technology and innovation
	b. Honesty and integrity
5. Hewlett Packard	a. Technical contributions
	b. Respect and opportunity for HP people
6. IBM	a. Full consideration for employee
	b. Make customers happy
7. Marriott	a. Friendly service and excellent value
	b. Continual self-improvement
8. Walmart	a. Provide value to customers
	b. Be in partnership with employees
9. Disney	a. No cynicism allowed
	b. Nurture American values
10. Procter & Gamble	a. Product excellence
	b. Honesty and fairness

Essentially all of the visionary companies experienced a "tipping point", to return to a theme in this chapter. The tipping was generally in revenue, profits or both. However, they were led by the management to a recovery strategy. So the issue was not whether a tipping point was in their future, but what to do when it appeared.

REFERENCES

1. Hall, C., "The Age of Synthesis", Peter Lang Publishers, 1995.
2. Blanchard, B. and Fabrycky, W., "Systems Engineering and Analysis", 5th Ed., Prentice-Hall, 2011.
3. Sherer, B. et al., "How To Do Systems Analysis", John Wiley, 2007.
4. Dan, G., see https://en.wikipedia.org/wiki/Daniel_Goldin
5. Simon, H., "The Science of Design", MIT Press, 1969.
6. Moore, S., "Supersize AI" (The 2.6 trillion transistor chip), IEEE Spectrum Magazine, July 2021.
7. "Developer Off-the-Self Systems" DOTSS, see Chapter 6.
8. Gladwell, M., "TheTipping Point", Little, Brown and Company, 2000.
9. Gladwell, M., "Blink", Little Brown and Company, 2005.
10. Gladwell, M., "Outliers", Little, Brown and Company, 2011.
11. Walden, David Walden, Garry Roedler, Kevin Forsberg, R. Douglas hmelin,, Thomas Shortell, INCOSE Handbook.
12. Peters, T. and Waterman, "In Search of Excellence", Harper & Row, 1982.
13. Collins, J, "From Good to Great", HarperBusiness, 2000.
14. Oppenheimer, R., see https://en.wikipedia.org/wiki/Robert_Oppenheimer
15. Gertner, J., "The Idea Factory – Bell Labs and the Great Age of American Innovation", Penguin Books, 2012.
16. Isaacson, W., "The Innovators", Simon and Schuster, 2014.
17. Duckworth, A., "Grit", Scribner, 2016.
18. Pirsig, R., "Zen and the Art of Motorcycle Maintenance", HarperCollins, 1974.
19. Edwards Deming, W., "Out of the Crisis", MIT Press, 2018.
20. Rechtin, E., "System Architecting", Prentice-Hall, 1991.
21. Hammer, M. and Champy, J., "Reengineering the Corporation", HarperBusiness, 1993.
22. Collins, J. and Porras, J., "Built to Last", HarperBusiness, 1994.

The Systems Approach and Thinking

5

Tomorrow's systems engineer has definitely not abandoned systems thinking. Indeed, the need remains and new ways are constantly being explored to get better and better at systems thinking. The benefits have not gone away – they have been supported by a real world need.

We begin with a reiteration of this author's elements of the systems approach [1]. This is provided in Table 5.1, and sets a baseline for systems thinking, which is listed as the last element.

From this point, we move on an example of systems thinking in the form of a meeting of the Association of American Railroads (AAR). The meeting was to consider strategic issues, and therefore, the following question was set forth:

what business are we in?

The resounding answer was – "we are in the railroading business, of course". The answer was not "we are in the transportation business", which, it is conjectured, is the reason that the railroad folks, by and large, never got into the air transport business. This is a revealing story, and can be applied to many strategic planning meetings and the failure to generalize the answer to the pivotal question – what business are we in?

From this construction, we move on to present the views of several researchers in the field. A researcher by the name of Leyla Acaroglu [2] has set forth what she calls tools for systems thinkers – Six Fundamental Concepts. The essence of this contribution is considered as follows:

1. *Interconnectedness*: The short form explanation of this element is that thinking shifts from linear to circular. It is a recognition that all system elements are interconnected.

DOI: 10.1201/9781003269274-5

TABLE 5.1   Elements of the systems approach [1]

1. Establish and follow a systematic and repeatable process
2. Assure interoperability and harmonious system operation
3. Be dedicated to the consideration of alternatives
4. Use iterations to refine and converge
5. Create a robust and slow-die system
6. Satisfy all agreed-upon user/customer requirements
7. Provide a cost-effective solution
8. Assure the system's sustainability
9. Utilize advanced technology, at appropriate levels of risk
10. Employ systems thinking

2. *Synthesis*: This refers to a process, that of putting pieces together to form larger, more relevant.
3. *Systems Mapping*: Shows relationship between elements of a system.
4. *Emergence*: As synthesis is achieved, emergent new properties become evident. These features are not accidents, they are new and often desirable features of the system as it evolves.
5. *Feedback Loops*: This feature is part and parcel of systems thinking. It is explicit rather than otherwise in a systems methodology known as system dynamics [3]. In general, there are two types of feedback Loops – reinforcing and balancing.
6. *Causality*: In this arena, we recognize the fact that some elements are actual causes for system behavior. For the AAR example, broadening the look from railroads to air transport allows them to see the present element (railroading) and also see the larger system (air transport). What a difference this broader view makes. What differences the system's view can lead to.

Ms. Acaroglu is also a developer and proponent of the Disruptive Design Method.

Railroads are a piece of, and element of, the larger concept of transportation. That perception represents the value of systems thinking, and why the systems engineer is in a domain different from the regular engineer.

To put it simply, *the system's view is the broader view*, and acknowledges that there is much to be gained from that view.

Here are some more examples along these lines:

- What business are we in, they are asking
- We're in the airport surveillance radar business, of course
- No we're not, we're in the "position location" business

The latter makes it clear that position location from space, and other types of radars (including sonars) do not escape our attention. If you're doing strategic planning, you'd best look out for serious omissions that could lead to extinction.

What business are we in, they asked?
We're in the film making and selling business, they answered.
NO we're not, we're in the image creation and manipulation business.
And so the very powerful film industry took a hit, every year, during
that 10 year period!

The system is the generalized entity; your current business may well just be a subset of that entity.

LIST OF DESCRIPTORS

In a study of the attributes of the successful systems engineer [4], the successful systems engineer had the following 10 associations:

1. A team player
2. Demonstrates continuous learning
3. Creative
4. Problem-solving
5. Analytic ability
6. Communication skills
7. Logical thinking
8. Attention to detail
9. Mathematical ability
10. Leadership

This author would elaborate upon the holistic theme with the following:

- Broad rather than narrow
- Integrated

- Expansive
- Generalized
- System wide
- Larger context
- Lateral
- Fusion
- Top level
- Inclusive

It is easy to obtain a "picture" of the successful systems engineer and systems thinker from these sets of descriptors.

Here again, scanning this list of the elements of the systems approach gives the reader a good idea as to useful thinking patterns while building a system.

Finally, a concise suggestion as to success factors for the productive systems engineer includes [5]:

1. Synthesizer
2. Listener
3. Curious/Systems Thinker
4. Manager/Leader
5. Expert/ESEP
6. Expert/Domain Knowledge
7. Perseverer

We note the fact that item (3) on this list refers directly to the Systems/ Thinker.

Systems thinking is so central to the field of systems engineering that we take extra time here to examine what has been written about systems thinking. The remainder of this chapter is devoted to this topic. The systems engineer of tomorrow must understand systems thinking and also be able to apply it on a regular basis as he or she is building systems.

Peter Senge

In his best-selling treatise [6], Senge identifies five disciplines that make for a learning organization, namely (1) personal mastery, (2) mental Models, (3) building a shared vision, (4) Team Learning, and (5) the fifth discipline (systems thinking). Senge asserts that the fifth discipline integrates the others, and that fifth discipline is systems thinking. So Senge was one of many that examined the significance of systems thinking and in a particular context.

INCOSE Handbook

INCOSE has addressed the various aspects of systems engineering in its fourth handbook [7]. INCOSE, in this handbook, confirms the significance of systems thinking in being able to execute systems engineering tasks and activities. The Handbook points to the following:

a. The role of system dynamics, from Forrester, in terms of understanding how systems work
b. The contributions of soft science and action research
c. The need to discover patterns
d. The behavior of the systems thinker and systems engineer

Russ Ackoff

A grandmaster of operations research and systems engineering, explains systems thinking in a short piece published by Pegasus [8]. Among his observations have been:

a. Improving the performance of parts of a system does not necessarily lead to improvements in the overall system
b. The best thing to do with many problems is not to solve them, but rather to dissolve them
c. Problems are not disciplinary in nature, but are holistic
d. The main function of many enterprises is not to maximize shareholder value, but rather to improve the lives of the managers in that enterprise

Peter Checkland

Professor Checkland [9] retired from the University at Lancaster in the UK where he established a forward looking research program. He is well known for his Soft Sysems Methodology (SSM) which is a process of inquiry that focuses on taking "action to improve". With respect to systems thinking, his key publication was the book "Systems Thinking, Systems Practice". Many folks in the UK consider Checkland's approach to be a real paradigm shift from "hard" to "soft" systems thinking.

J. Boardman and B. Sauser [10]

These authors have introduced "systemic" thinking which is a framework and graphical procedure for thinking and solving problems. They show how systemigrams have been used to explore the behavior of systems and building maps of systems' interactions and Worlds of Systems. These same two authors also published a book with the title "Systems Thinking" [11]. This latter text emphasizes opportunities as well as difficulties in carrying out a systems approach. They "place a high value on the thinking process" in order to successfully solve problems.

Michael Jackson [12]

Jackson notes that there are two kinds of systems thinking: hard and soft. He points to Checkland, also from the UK, as having formulated the Soft Systems Methodology, whereas he focuses on creative holism. Key topics that he explores are system dynamics, interactive planning, systems heuristics, organizational cybernetics, and total systems intervention.

Donella Meadows

Ms. Meadows was a proponent of the Club of Rome philosophy and field of study [13]. She was the lead author of Limits to Growth and leaned heavily upon system dynamics as defined and developed by Forrester [3]. She defined a system as composed of three things:

    a. Elements
    b. Interconnections, and
    c. Functions

and explored in some detail each of them. She explained the relationship between system structure and its behavior, strongly supported by quantitative analysis

Jamshid Gharajedaghi

In his second edition, this author [14] provides us with a simple statement: "systems thinking is the art of simplifying complexity". He also sets forth the four foundations of systems thinking which he takes to be:

a. holistic thinking
b. the dynamics of multiloop feedback systems
c. self-organization and movement toward a predefined order
d. interactive design

Gerald Weinberg

This author was a prolific researcher and writer, with his popular treatise on General Systems Theory taking a leading position [15]. Weinberg approaches this dense topic from a "problem-solving" point of view, using these three steps:

1. state the problem
2. make observations regarding the problem
3. formulate solution

This author poses several "laws" (like the eye-brain law) and several principles (like the principle of invariance). This is an excellent exposition, but perhaps an unconventional one for the topic of systems thinking.

S. G. Haines

Mr. Haines called his treatise "Systems Thinking and Learning" [16]. It started with a definition of systems thinking as "a world view and way of thinking whereby we see the entity as a whole, with its fit and relationship to its environment as primary concerns". It featured the A-B-C-D- model where A = Output, B = Feedback Loop, C = Input, and D = Throughput. This is the classical feedback control graphic, with well-defined behavior patterns. The Haines approach involves moving from chaos and complexity to elegant simplicity.

Virginia Anderson and Lauren Johnson [17]

These authors have provided an overview of systems thinking, with an emphasis on tools, behavior over time graphs and causal loop diagrams. These perspectives are common to other treatments of systems thinking, as readers prefer diagrams for more immediate understanding.

Daniel Kim

Mr. Kim points out that systems thinking differs markedly from the more conventional reductionist way of solving problems [18]. His treatise suggests that it is possible to give the reader the "language and tools to apply systems thinking principles and practices to your own organization".

L. B. Sweeney

Ms. Sweeney starts her text by rejecting analysis [19], and setting forth the following:

    a. seeing the world in terms of wholes (vs. individual parts)
    b. seeing how these parts interact together
    c. seeing overall system relationships and behavior
    d. seeing how our mental models may need to change
    e. seeing patterns of causality

Barry Richmond

Mr. Richmond addresses systems thinking by looking at the problem-solving dynamic of the following steps [20]:

    a. Specify the problem or issue under consideration
    b. Construct a hypothesis
    c. Test the hypothesis
    d. Implement changes
    e. Communicate overall understanding

He also relies on various models that lead to seven critical thinking skills and an appropriate set of answers to more-or-less any problem set.

Michael McCurley

Mr. McCurley relates systems thinking to the old but highly relevant text of the Tao Te Cing [21]. Other themes explored are:

    a. Balance
    b. Simplicity

c. Models
d. World and system dynamics

He also acknowledges the most significant work by Jay Forrester.

John Gall

Dr. Gall, in addition to being an author, is a retired Pediatrician. In his treatment of children for over 40 years, he looked at generalized problems and constructed a series of "Laws" in his overall critique of systems theory [22]. Here are some of his "Laws":

a. simple system may or may not work
b. complex system that works is likely to have evolved from a simple system that worked
c. a complex system designed from scratch never works

It seems that there is a strong connection between Gall's Laws and agile software development. This is to be noted in the section on software and agile development.

O'Connor and McDermott

These authors [23] emphasize the following points with respect to systems thinking:

a. Feedback loops are the essence of systems thinking
b. Time delays are significant in terms of systems' behaviors
c. A system works as well only as its weakest link
d. Results, in general, are not proportional to applied effort

Looking at the first of these citations regarding feedback loops, this author notes that feedback loops have been treated in detail in texts on feedback control systems and servomechanisms.

Albert Rutherford

Rutherford approaches complexity in systems by addressing nine "archetypes" that explain behavior and suggest new ways of defining and dealing with problems. His book on Systems Thinking [24] has the following subtitle: "Use System Archetypes to Understand, Manage and fix Complex Problems and Make Smarter Decisions". If we believe Rutherford, this book should be on every systems engineer's bookcase.

Whitcomb et al.

"Systems Thinking" [25], this offering is a set of some 12 articles, as compiled by three editors. Topics of interest include the following:

a. Applications to systems science and system dynamics
b. Approaches to problem areas
c. Models (e.g., maturity)
d. Theoretical frameworks
e. An architecture for systemology

Marcus Dawson

Dawson [26] explores a variety of topics in some detail, including:

a. The contrast between linear and lateral thinking
b. Mental models and paradigms
c. Keeping one's focus
d. Chaos theory
e. System archetypes
f. The iceberg model

R. Ackoff and J. Gharajedaghi [27]

This is largely a review of some 40 f-laws (flaws) from Ackoff, with a foreword from the latter author. They are oriented toward business issues, as in systems thinking for organizations and managers. Professor Ackoff was also the author of "reflections on systems and models" and more than 200 papers.

Michael Goodman

Mr. Goodman, writing in The Systems Thinker, gives advice to the "beginner" [28]:

a. Investigate and use the archetypes
b. Practice every day
c. Use at work and at home
d. Use the tools, to include feedback causal loops and diagnostics
e. Reject inappropriate mental models

F. Capra and P. L. Luisi [29]

"The Systems View of Life – A Unifying Vision", Cambridge University Press, 2014. These authors attempt to integrate ideas, theories, and models so as to produce an overall framework which they call a systems view of life. Areas explored include the following:

a. Dissipative structures
b. Social networks
c. Autopoiesis
d. evolution

Monat and Gannon 1 [30]

"Using Systems Thinking to Solve Real World problems", College Publications (UK), 2017. These authors have looked at the holistic nature of systems, emphasizing causal loops and diagrams, behavior over time graphs, dynamic modeling, archetypes, and root cause analysis. They have been especially interested in how these concepts apply to problem-solving, in distinction to formulating new theory.

Monat and Gannon 2 [31]

"What is Systems Thinking? A Review of Selected Literature Plus Recommendations. This very useful contribution provided a complete review of the systems thinking literature, including several pieces that this author might otherwise have overlooked.

RosalindArmson [32]

"Growing Wings on the Way: Systems Thinking for Messy Situations", Triarchy Press, 2011. Ms. Armson defines what she calls "messy situations" and proceeds to explore how systems thinking leads one to ways and means of "cleaning up" these situations. It is a primer on applications of systems thinking to a variety of problem areas, written clearly and presented imaginatively.

As we close this chapter, I (this author) recall an incident that illustrates what systems thinking looked like when I (the author) was a young project manager. I was running a modest software project that had a specific clock time requirement. As we proceeded into our early system testing, we found that we were not meeting the clock time requirement. I pushed harder on the brainpower of the team, hoping that someone could figure out how to meet this requirement. We were failing, in this respect, when the youngest member of our team came up with the following:

> why don't we go back to the government manager, explain what was happening and ask for relief on this requirement?

We discussed this approach very carefully, and then finally put together a long explanation of what we had done and why, and made an appointment to see the government manager. We told our story, and finally the government person got up and declared something like:

> You've made a good case for relief, and given the progress you've made and the more than adequate performance of your system, I'm going to grant you relief from this particular clock time requirement.

Our team was very happy with this conclusion, and then mindful of the fact that the government manager was part of the system, and a happy part at that. And all of us were more than pleased to be reminded that the government manager was indeed part of the system.

REFERENCES

1. Eisner, H., "Systems Engineering – Fifty Lessons Learned", CRC Press, 2020.
2. Leyla, A., see https://en.wikipedia.org/wiki/leyla_acaroglu
3. Forrester, J., Principles of Systems, Pegasus Communications, 1968.
4. McClements, D., "Ten Characteristics of Successful Engineers", see Newengineer. com, January 8, 2019.
5. Eisner, H., "What Makes the Systems Engineer Successful?", CRC Press, 2021.
6. Senge, P., "The Fifth Discipline - The Art and Practice of the Learning Organization", Currency, 1990.
7. Walden et al., INCOSE Handbook, 4th Ed., John Wiley, 2015.
8. Ackoff, R., "A Lifetime of Systems Thinking", Pegasus Communications.

9. Checkland, P., "Systems Thinking, Systems Practice", John Wiley, 1981.
10. Boardman and Saucer, "Systemic Thinking", John Wiley, 2013.
11. Boardman and Saucer, "Systems Thinking – Coping With 21st Century Problems", CRC Press, 2008.
12. Michael, J., Creative Holism for Managers, Wiley, 2003.
13. Meadows, D., Thinking in Systems, Earthscan and Chelsea Green, 2009.
14. Gharajedaghi, J., "Systems Thinking - Managing Chaos and Complexity", Elsevier, 2006.
15. Weinberg, G., "An Introduction to General Systems Thinking", Dorset House, Vol. 1975, 2001.
16. Haines, S. G., "Systems Thinking and Learning", HRD Press, 1998.
17. Anderson, V. and Johnson, L., "Systems Thinking Basics, Pegasus Communications, 1997.
18. Kim, D., "Introduction to Systems Thinking", see https://the systems thinker.com
19. Sweeney, L. B., "Systems thinking: A Means to understand Our Complex World", Pegasus Communications, 1995.
20. Richmond, B., "The Thinking in Systems Thinking: How Can We Make It Easier to Master?", The Systems Thinker.
21. McCurley, M., "The Tao of Systems Thinking", Three to the Fourth Power, 2015.
22. Gall, J., "General Semantics: An Essay on How Systems Work, and Especially How They Fail", The New York Times Book Company, 1975.
23. O'Connor and McDermott, I., "The Art of Systems Thinking", HarperCollins, 1997.
24. Rutherford, A., "The Elements of Thinking in Systems", Kindle Direct, 2019.
25. Whitcomb, et al., "Systems Thinking", MDPI, 2020.
26. Dawson, M., "Thinking in Systems and Mental Models", 2020.
27. Ackoff, R. and Gharajedaghi, J., "Systems Thinking for Curious Managers: With 40 New Management f-laws", Triarchy Press, 2010.
28. Goodman, M., "Systems Thinking – What, Why, When, Where and How?", The Systems Thinker.
29. Capra, F. and Luisi, P. L., "The Systems View of Life – A Unifying Vision", Cambridge University Press, 2014.
30. Monat and Gannon 1, "Using Systems Thinking to Solve Real World Problems", College Publications, UK, 2017.
31. Monat and Gannon 2, "What Is Systems Thinking? A Review of Selected Literature Plus Recommendations", American Journal of Systems Science, 2015.
32. Rosalind, A., "Growing Wings on the Way: Systems Thinking for Messy Situations", Triarchy Press, 2011.

Architecting and Modeling

6

This is the first of three chapters dealing with "long poles in the tent". There are four areas explored in this chapter, namely:

1. Architecting
2. Modular Design
3. Modeling and simulation (M & S)
4. Model-based systems engineering (MBSE)

ARCHITECTING

This author considers architecting the most important skill area that needs to be executed by the systems engineering team. This author sees the possibility of high-performance teams carrying out the task of architecting the system in question. In all likelihood, these teams will accept advanced versions of DoDAF, the Department of Defense Architectural Framework [1]. Why is this the case? The DoD has taken the time and put in the energy to advance this approach and are likely to continue to do so. However, this is a flawed approach, for the reasons discussed below.

The DoDAF defines an architecture as follows:

> the structure of components, their relationships, and the principles and guidelines governing their design and evolution over time

The DoDAF architecture approach continues with the notion that it is important to define three views, namely:

a. The operational view (OV)
b. The systems view (SV)
c. The technical standards view (TV)

DOI: 10.1201/9781003269274-6

with each being defined as below [2], with the precise original words:

- "The operational view is a description of the tasks and activities, operational elements, and information exchanges required to accomplish DoD missions", which "include both warfighting and business processes"
- "The SV is a set of graphical and textual products that describe systems and interconnections providing for, or supporting DoD functions", which "include both warfighting and business functions"
- "the TV is the minimal set of rules governing the arrangement, interaction, and interdependence of system parts or elements. Its purpose is to ensure that a system satisfies a specified set of operational requirements"

The DoDAF also defines architectural aspects that concern all views, and these are captured in the All-Views (AV) products.

Proceeding forward, the DoDAF articulates *essential* views, as follows:

AV-1 Overview and Summary Information
AV-2 Integrated Dictionary
OV-1 High Level Operational Graphic Concept
OV-2 Operational Node Connectivity Description
OV-3 Operational Information Exchange Matrix
SV-1 System Interface Description
TV-1 Technical Architecture Profile

Despite the specificity of the above, this author defines an architecture in a particular way [3], as follows:

an architecture is an organized top-down selection and description of design choices for all the system functions and sub-functions, placed in a context to ensure interoperability and satisfaction of final system requirements

We see in the above definition a way to proceed with architecting a system. That way is to first define the functions to be carried out by the system, otherwise known as functional decomposition. After that, the architect sets forth the design approaches for each function. Using top-level logic and the approach known as Analysis of Alternatives [4], the architect defines three approaches to instantiating each function. These three approaches are attempts to synthesize (a) a low cost system, (b) a knee-of-the-curve system, and (c) a high-effectiveness system. An example of this approach is shown in Figure 6.1, for a house [5].

FIGURE 6.1   Synthesis step for architecting a house

FUNCTIONS IN HOUSE	LOW COST	KNEE-OF-CURVE	HIGH EFFECTIVENESS
Overall style	Ranch	Faux farm	Contemporary
Food preparation	Standard kitchen	Extra built-ins	Island + extra counters
Sleeping facilities	Three bedrooms	Four bedrooms	Five bedrooms + den
Environment	Single furnace/AC unit	Two furnace/AC unit, medium capacity	Three zone furnace/AC high capacity
Stairwells	Flat, no stairs	One seven step	Two seven step
Recreation	Small deck	Add small rec room, larger deck, add brick BBQ	Wrap around deck, add warmers, outdoor shower
Bathing	Standard toilet/ bath	Add bidet + two sinks	Add cabinets and Jet tub/bath
Space/size	3,500 sq ft	4,500 sq ft	6,000 sq ft
Lawn/garden	Small lawn	Buried water lines	Large lawn + gazebo + buried water lines
Living/dining spaces	Standard	Larger spaces	High ceilings
Auto facility	One-car garage	Two-car garage	Three-car garage, built in shelving
Safety/security	No extras	Camera, tape, alarm	Add internet connection
Built Ins	Standard	Add display, dining	Display into living room/den
Plumbing	Standard	w/ Multiple Flareouts	Add sprinkler system
Electrical	Copper standard amps	Add 50%	Add 100% for growth
Special amenities	Standard closets/ lights	Extra room for office, bookcases	Extra closets, bookcases, lighting, room for movies, add elevator

EVALUATION OF ALTERNATIVE ARCHITECTURES

The architecting process continues with a formal evaluation of the three alternatives. This is done on a cost-effectiveness basis, using procedures from that discipline. An example of such an evaluation follows.

The next step is to define a set of evaluation criteria, and assess the cost-effectiveness of each alternative. This is displayed in Figure 6.2 below.

Each architecture is scored on a scale of 1–10, against each criterion. Then the effectiveness is calculated as the product of the score times the weight. The bottom line effectiveness measures are then 6.2, 7.2, and 8.95, respectively. The costs are $700 K, $1.2 M, and $2.5 M, respectively. The analyst is looking for the most cost-effective architecture. In general, we may make the following observations.

If money is an issue, one might select the $700 K home, even though many features are not present. If money is not at all an issue, one might say – I have the money and I like the score as well as the various features that go along with the contemporary style house. Finally, one might conclude that the knee-of the curve house is a good compromise. Further, I don't really need all the features, given my current circumstances.

The further observation is that there are four steps that we wish to execute in this architecting approach:

1. Functional decomposition
2. Synthesis of three alternatives

FIGURE 6.2

EVALUATION CRITERIA	WEIGHTS	LOW COST		KNEE-OF-CURVE		HIGH-EFFECTIVENESS	
		SCORE	WXS	SCORE	WXS	SCORE	WXS
Performance	20%	6	1.2	7	1.4	9	1.8
Maintainability	25%	5	1.25	6	1.5	8	2.0
Reliability	10%	7	0.7	8	0.8	9	0.9
Holds Value	10%	6	0.6	7	0.7	9	0.9
Risk	15%	7	1.05	8	1.2	9	1.35
Feels Like Home	20%	7	1.4	8	1.6	10	2.0
SUMS	100		6.2		7.2		8.95
COSTS			$700,000		$1,200,000		$2,500,000

3. Evaluation of alternatives (analysis)
4. Selection of most cost-effective solution

These steps are quite different from the DoDAF approach, and conclude with cost and effectiveness considerations, a familiar approach for many architects and system evaluators.

Modular Design

In conjunction with the above method of architecting, there is an opportunity to move forward with modular design. We explore some elements of modular design in this section of the chapter. This author expects that modular design will make very significant progress in the next couple of decades.

Modular Design involves well-defined modules that fulfill, or satisfy, a given function or sub-function, re-usable with standard interfaces. It is especially compatible with the architecting approach described above since the synthesis step requires the instantiation of functions and sub-functions.

One concept regarding modular design that is set forth by DDR&E is called Modular Open Systems Approach (MOSA) [6]. MOSA is presented as part of the Defense Standardization Program [7] which is defined as "a technical and business strategy for defining an affordable and adaptable system". Through the use of "architecture modularity, open system standards, and appropriate business practices". The DoD believes that these kinds of benefits [8] will accrue from widespread use of MOSA:

a. Cost savings or cost avoidance
b. Schedule reduction
c. More opportunities for technical upgrades
d. Greater interoperability

Modular Design holds out great promise for achieving what Dan Goldin was advocating, namely, faster, better and cheaper results in terms of building systems. And that has to be one of the goals of the future.

Modeling and Simulation (M & S)

This is selected as a key area of the future for systems engineering. The baseline concept is that we are developing a system, or a system of systems (SoS), and we wish to explore its features and performance. So, instead of working with the real system, which may not be more than a paper design, we build and work with a model or a simulation of the system. Building a model of a large-scale system is an interesting as well as difficult task. This is illustrated by the two examples below.

THE NATIONAL AVIATION SYSTEM (NAS)

This author bid on and received a contract from the FAA to construct a National Aviation System (NAS) Model [9]. As the work evolved, it became clear that the over-arching model should have several sub-models, to include the following:

  a. Airport capacity
  b. Airspace capacity
  c. Demand
  d. Trip Time
  e. Delay
  f. ATC Availability
  g. Service Availability
  h. Landing System Performance
  i. Pollution
  j. Noise
  k. Energy Utilization
  l. Safety
  m. Security

Considering the above, it turned out that airport capacity was the most simple model, using FAA documentation. The most difficult was the airspace model. So one concept relative to the modeling part of the M & S grouping is to build a model of the system in question or the SoS when the situation becomes substantially complex. For a SoS, the top-level model is likely to be conceived as a "model of models".

MODELING SYSTEMS OF SYSTEMS

Y. Y. Haimes sets forth a method for modeling SoS [10] that he calls "Phantom System Modeling". This method focuses on systems with multiple functions, operations and stakeholders. It models inter- and intradependencies between subsystems of a complex system. In that sense it deals with complexity quite directly. It also deals with hierarchies, Bayesian models, and holographic representations.

This author uses his parameter dependency diagramming for modeling SoS [11]. This method explicitly shows the dependencies between key parameters of a system or systems. This is done in graphical form so that the dependencies are clear and obvious. These diagrams also stop short of describing the relationship between parameters. This method turned out to be quite useful in formulating the NAS model.

THE BASIC POINT DEFENSE MISSILE SYSTEM (BPDMS)

Yet another example of the use of a model is the contract formulated by this author with respect to the Basic Point Defense Missile System (BPDMS). This contract effort was for the Navy base known as the Naval Surface Warfare Center, at Port Hueneme in California. Here the commanding officer wanted to assist the staff by building a model of the essential performance of the BPDMS. With such a model, actions contemplated by the staff could be tested using the model instead of the real system.

Strategic Defense Initiative (SDI)

President Reagan, back around 1981, supported what was known as the Strategic Defense Initiative (SDI) program. Contractors scrambled to get a foothold in the program, looking carefully at what they had to offer by way of contributing to the program. This author was no different and recognized that a preliminary simulation of the SDI scenario was built by a particular Vice President of the company. This capability was presented to a large and well-known defense company, and they immediately offered a contract to extend and improve the simulation and present ongoing results to an eager government customer. The SDI simulation gave the user the capability of placing x satellites in y orbits, and then placing weapons (e.g., kinetic kill and high-energy laser) aboard these satellites. The simulation called for firing these weapons against enemy missiles that were launched to do damage to the United States. Using this simulation, the user could then explore, in considerable detail, at least the following:

a. Various firing strategies
b. Various command and control approaches
c. Weapon system performance against various threats
d. Various satellite payloads
e. Adding satellites and adding orbital planes

It was then possible to explore SDI performance on the computer instead of the real system, which was not yet built, nor in many cases, a conceptual system.

A follow-on to the SDI program has been the national missile defense program. This program has the following elements:

a. Ground-based interceptor missiles
b. The Aegis Ballistic Missile Defense System
c. Terminal High-Altitude Area Defense
d. Airborne Systems

SIMULATION SOFTWARE

There are many simulation packages available in today's market. This availability supports the value of simulation inn terms of exploring the design and performance of a new system. A company by the name of Capterra has provided an annotated list of such software, as shown in Table 6.1. That is followed by Table 6.2 which has another list, captured via an internet search. The interested reader should proceed with these lists and other key word searches.

TABLE 6.1   Simulation software suggested by Capterra

- FlexSim
- iGrafx
- Simcad Pro
- SimSolid
- OnScale Solve
- Solid Edge
- MATLAB
- AnyLogic
- Unreal Engine
- Fusion 360
- SimScale

TABLE 6.2   Simulation software provided by internet search

- Solidworks
- Navisworks
- Teamcenter
- Autodesk Fusion 360
- Solid Edge
- PTC Creo
- SimScale
- Simulink
- ETAP PS
- Altair Inspire
- EDA Custom IC
- MSC Apex, One

We note the fact that both lists contain the packages Solid Edge, Fusion 360, and SimScale.

For those old timers that have had some contact with commercial simulation packages, the following list is appropriate [11]:

- SIMLAB
- SIMAN
- GPSS
- DYNAMO
- SIMSCRIPT
- SLAM
- GASP

MODEL-BASED SYSTEMS ENGINEERING (MBSE)

MBSE was the apparent brainchild of Wayne Wymore [12], a Professor at the University of Arizona, and a champion of the "new" field of systems engineering. A more recent champion of MBSE has been Sandy Friedenthal of Lockheed Martin [13]. His book on SysML is a classic in that it exemplifies a specific approach to MBSE. SysML is a systems modeling language that has achieved wide acceptance, which is not to say that this is the only approach set forth by others. As an example, Dov Dori has pointed us to OPM [14], the object-process methodology. Estefan has carried out a survey of MBSE Methodologies [15]. And Ramos has cited MBSE as an emerging approach for modern systems [16]. Other examples are IBM's Engineering Lifecycle Optimization [17] and thyssenkrupp Marine Systems MBSE [18]. The latter is viewed from the perspective of reducing complexity, a benefit of some approaches to MBSE.

Innoslate

An especially notable example of MBSE is called innoslate [19] and is provided by a company named SPECInnovation and its president, Steven Dam. According to data provided by Dr. Dam, Innoslate "uses the LML (Lifecycle Modeling Language) and the SysML to describe the concepts and diagrams". The LML addresses systems engineering and project management matters. The SysML is an extension of a subset of UML. This is an excellent example of how MBSE has proceeded and is being made available to the systems engineering community. One can only imagine where Innoslate and its derivatives will be in a couple of decades. It will certainly be mandatory to have tomorrow's systems engineers fully conversant with a tool such as Innoslate or similar systems.

There is little doubt that MBSE must be part and parcel of the tools needed by the systems engineering community, moving into the future. Although SySML is an early version of MBSE, many new ideas and versions are likely to be developed and available to this community. Some of these are also likely to come out of the general M & S world, using a commercially available package.

REFERENCES

1. DoDAF, v. 2.02, Department of Defense Architectural Framework.
2. DoDAF, v. 1.0, Department of Defense Architectural Framework, Vol. 1, February 9, 2004.
3. Eisner, H., "Systems Architecting – Methods and Examples", CRC Press, 2020.
4. "An AoA (Analysis of Alternatives) Handbook – A Practical Guide to the AoA", Office of Aerospace Studies, Kirtland Air Force Base, NW, July 6, 2016.

5. Eisner, H., "Systems Architecting – Methods and Examples", CRC Press, 2020, p. 26.
6. MOSA 1, DDR&E; see https://ac.cto.mil>mosa
7. MOSA 2, see dsp.dla.mil
8. MOSA 3, see MITRE; see https://aida.mitre.org
9. "National Aviation System Model – Overview", March 18, 1977, ORI, Inc., 1400 Spring Street, Silver Spring, MD 20910.
10. Haimes, Y., "Modeling Complex Systems of Systems With Phantom System Models", see https://onlinelibrary.wiley.com, May 16, 2012.
11. Eisner, H., "Computer-Aided Systems Engineering", Prentice-Hall, 1988.
12. Wymore, W., "Model-Based Systems Engineering", CRC Press, 1993.
13. Friedenthal, S., Moore, A. and Steiner, R., "A Practical Gide to SysML: The Systems Modeling Language", Elsevier, 2015.
14. Dori, D. "MBSE with OPM and SysML", see www.amazon.com
15. Estefan, J., "Survey of Model-Based Systems Engineering Methodologies", INCOSE MBSE Initiative, see https://core.ac.uk
16. Ramos, A., "Model-Based Systems Engineering: An Emerging Approach for Modern Systems", IEEE Transactions on Systems, Man and Cybernetics, January 2012.
17. IBM Engineering Optimization – Engineering Insights, see https://www.ibm.com
18. "Simplifying Complexity Through MBSE", thyssenkrupp Marine Systems, Dassault.
19. Innoslate 101, see https://help.innoslate.com

Software Reuse and Agility

7

This is the second chapter that deals with "long poles in the tent". There are four such "poles", namely:

1. Developer-off-the-self-systems (DOTSS) – Reuse of Software
2. Agile development
3. Software architecting
4. Resilience

The reader is reminded that these "long poles" are areas of special importance to the systems engineer and how he or she is able to perform some 20 years down the road from where we are today.

DOTSS

DOTSS refers to an approach to system development that involves the reuse of entire systems. It considers non-complex systems only, and excludes high-performance command and control systems. It generally also excludes what are called real time systems, requiring very rapid timelines. It includes:

a. Human resources tracking systems
b. Non real-time legislative tracking
c. What might be called vanilla systems, with relatively slow performance envelopes. These also will be called non-complex IT systems

DOI: 10.1201/9781003269274-7

The primary research into this concept was carried out by this author over a period of several years. The initial paper with respect to DOTSS set forth a procedure leading to being able to develop a system at one-fifth of the time, and also one-fifth of the costs. Taken together, this represents one twenty-fifth of the total effort. That is equivalent to a 2,500% improvement, an astonishing number. In order to demonstrate this point we go back to a statement made by Dan Goldin, a previous administrator of NASA. When he became the head of NASA, he made the declaration that he was going to try to develop systems faster, cheaper, and better. Although he was not able to demonstrably achieve this, apparently he did accept it as a goal.

SOFTWARE REUSE

In conjunction with the above DOTSS approach which promises very large improvements by re-using entire systems, this author is suggesting a major new initiative addressing the matter of software reuse. In that connection, we will briefly look at a book source dealing with software reuse [2] and a more recent Wikipedia treatment [3].

In a comprehensive examination of software reuse, Leach explores such topics as:

a. Domain analysis
b. Standard interfaces
c. Reuse libraries
d. Certification of software components
e. The economics of software reuse

Although tis author stops short of citing substantial savings through software reuse, he suggests several approaches that are worthy of consideration.

The Wikipedia treatment includes a reference to work from Bell Labs in 1968 in the context of a proposed "basing the software industry on reusable components" [3]. Other relevant topics in this regard include:

a. Software libraries
b. Opportunistic and planned reuse
c. Systematic reuse
d. Components
e. Reuse metrics

Software reuse remains one of the key initiatives suggested by this author, preparing the systems and software engineer for significant improvements in overall productivity.

AGILE DEVELOPMENT

An agile characteristic can be viewed as a quality of maintaining agile processes that allow an enterprise to respond quickly to changing business needs. If the response is rapid and pointed at the need, then the enterprise will grow stronger with each such response.

Agile has often been applied to software development where there has been a great need. And "agile" has also been placed in the context of an overall acquisition strategy [4]. MITRE has defined four core elements for agile acquisition:

   a. A program structure with small, frequent capability releases
   b. Placing value on software development itself over pure documentation
   c. Being maximally responsive to changes in operations, technology, and budgets, and
   d. Active involvement of users of the system in question

MITRE goes on to describe release, sprint, and daily scrum as necessary in an agile program and as follows:

   a. A release is a delivered capability resulting from multiple sprints
   b. A sprint is a rapid high-priority development, that is fully tested, and
   c. A daily scrum is a team meeting that fully and quickly addresses progress and ways to solve any problems that have arisen

MITRE proceeds with suggestions as to when to use agile. Here are some of the situations in which such use is appropriate:

   a. When a full set of requirements cannot be defined up front
   b. When developed requirements can be decomposed into small iterative tasks
   c. When users of the system can be made available throughout the system's development
   d. When a full collaborative environment between users, program staff, and developers can be established

Beyond the MITRE approach, the literature provides a set of principles for agile software development [5], as paraphrased below:

a. One achieves customer satisfaction through early and continuous delivery of software that works
b. One welcomes requirements, even those formulated late in the program
c. Software is delivered frequently (weeks vs. months)
d. Close involvement between business folks and developers
e. Projects undertaken by motivated people who are trusted to get the work done
f. Co-location between developers and users
g. Delivery of working software is only test of "system"
h. Continuous attention to excellence in software design and delivery
i. Simplicity of design and approach
j. Use of self-organizing responsible and high-performance teams
k. Continuous checking, by the team, as to how to be maximally effective

The perspectives regarding agile have been based, in general, upon an "agile manifesto" that was formulated in 2001 [6]. The motivation had to do with attempting to improve the way in which software was developed. That need has not gone away. The DOTSS approach, discussed above, is one in which improvements in software systems can be achieved through the notion of software reuse.

SOFTWARE ARCHITECTING

As architecting is considered to be a crucial skill for the future system engineer, it is especially true in the software arena. There are several cogent sources to consult when addressing the matter of software architecting. As examples, see the reference in this connection [7]. A good definition of software architecture can readily be found, as below [8]:

Architecture is "the fundamental organization of a system, embodied in its components, their relationship to each other and the environment, and the principles governing its design and its evolution"

However, a clear concise roadmap for developing a software architecture has yet to be articulated. So this area is one of the key capabilities for the future systems engineer. The reader is further pointed to tis author's text dealing with systems architecting [9]. This text suggests an approach similar to that for systems engineering.

RESILIENCE

It will be recalled that the "systems approach", as defined by this author [10], includes designing a system such that it is "slow-die"; or in other words, no single failure will kill the mission. This is a far-reaching "requirement", but essential in its implementation. It is especially relevant to manned missions when the stakes are so high. But it also holds for non-manned missions, as the example below suggests.

Tis author, back in the 60s, was advisor to a team building the weather satellite known as Nimbus. This was a three-axis stabilized system with solar panels driven by special motors. As it turned out, the solar panel drive motor failed, meaning that the satellite could not get enough solar energy to sustain its mission. This "single point" failure killed the satellite and gave us all something to think about, and also act upon. This author resolved that for the future, to the maximum extent possible, he would not allow a single point failure that would result in a mission failure. Single point problems would be backed up with redundancy, wherever and whenever possible.

Another way to look at this type of problem is to build resilience into the system. And for a cogent discussion of resilience, we look at the work of Yakov Haimes [11]. He presents the following definition:

> resilience is the ability of a system to withstand a major disruption within acceptable degradation parameters and to recover within an acceptable time and composite costs and risks

We note that this definition places resilience in the context of risk, and also composite costs. Not noted is the notion that it is the responsibility of the systems engineer to address the matter of resilience, as well as how to measure it and how to assure that it is a part of the systems in question. A metric for resilience would be welcome as this author has not yet found an acceptable one.

REFERENCES

1. Eisner, H., "Reengineering the Software Acquisition Process Using Developer-Off-the-Shelf Systems (DOTSS)", 1995 IEEE International Conference on Systems, Man and Cybernetics, October 22, 1995.
2. Leach, R. J., "Software Reuse", McGraw-Hill, 1997.

3. "Code Reuse", see https://en.wikipedia.org/wiki/Code_reuse
4. See "Agile Acquisition Strategy", The MITRE Corporation.
5. See https://en.wikipedia.org/wiki/Agile_software_development
6. Agile manifesto; see https://agilemanifesto.org
7. Garlan, D., "Software Architecture", Encyclopedia of Software Engineering, 2nd Ed., John Marciniak (E-I-C), John Wiley, 2002, p. 1318.
8. Maier, M., "System and Software Architecture Reconciliation", Systems Engineering, Vol. 9, Issue 2, Summer, 2006.
9. Eisner, H., "Systems Architecting – Methods and Examples", CRC Press, 2020.
10. Eisner, H., "Systems Thinking – A Guide to Systems Engineering Problem Solving", CRC Press, 2019.
11. Haimes, Y., "On the Definition of Resilience in Systems", Risk Analysis, Vol. 29, Number 4, Wiley Online Library, 2009.

Data and Decisions

8

This is the third "long poles" chapter, dealing with the following topics:

a. Digital engineering
b. Data and information management
c. Security/trust/risk
d. Decision support frameworks
e. Big data and analytics

DIGITAL ENGINEERING

There appears to be a lot of current support for research and development in the field of digital engineering. This support is likely to have originated in the OSD, as part of the DoD [1]. DDR&E has stated that digital engineering is critical in an environment of increasing global challenges as well as dynamic threats. DDR&E goes on to say that digital engineering is an update of traditional systems engineering approaches through the use of computational technology, modeling, analytics, and data sciences. Moving aggressively in this area, DDR&E tasked the SERC at the Stevens Institute to develop a digital engineering competency framework review procedure [2]. The latter (DECF – v. 1.1) consists of five elements, as follows:

a. Systems software
b. Modeling and simulation
c. A digital enterprise environment
d. Digital engineering and analytics, and
e. Data engineering

DOI: 10.1201/9781003269274-8

In June 2018, the DOD published a digital engineering strategy [3]. The digital engineering initiatives cited in that report are articulated under four categories:

1. Policy/guidance
2. Pilots
3. Implementation
4. Tools

Key items cited in the above include:

- Setting up cross-service virtual environment
- Assure service level implementation plans
- Convene tool summit
- Sponsor workshops in digital engineering

Some of the benefits that the DoD saw at that time were:

a. Better transparency and decision-making
b. Greater adaptability
c. Increased efficiency in engineering and acquisition practices

INCOSE's Vision 2035 includes a special place for digital engineering [4]. In particular they have envisioned several impacts of the digital transformation, taking it for granted that the transformation will and must occur. They claim that the transformation will change products, work, and decision-making in significant ways. A quote from the INCOSE Vision that is relevant is:

> advancing digital technologies are enabling model-based systems engineering (MBSE) practices, but emerging computational capabilities, cloud infrastructure, and data and information discovery are not being fully leveraged compared to other engineering and scientific disciplines.

DATA AND INFORMATION MANAGEMENT [5]

The key agency responsible for data and information management is the Defense Information Systems Agency (DISA). This agency has set forth the following three goal areas:

a. Operate and defend
b. Adopt before we buy and buy before we create, and
c. Enable people and reform the agency

Under "operate and defend", the agency has articulated the following five areas:

1. Modernize the infrastructure
2. End user support
3. Computing
4. Defensive cyber operations
5. Readiness

Obviously, a great deal of thought has gone into setting up these strategic operations in a world that has become more complex each year. These goals are enabling DISA to conduct DoDIN operations for the joint warfighter, in any and all domains. The area known as defensive cyber operations is especially challenging and worthy of a lot of systems engineering attention in the next 20 years.

DISA, in 2021, has seen the need to define a new strategy [5]. This involves bringing together five "not so easy" pieces they have called lines of effort, namely:

1. Command and control
2. Leveraging innovation
3. Data as a center of gravity
4. Cybersecurity
5. Empowering the workforce

This latter item is a theme (and a problem) with many government agencies going forward.

SECURITY/TRUST/RISK

These three elements are critical to handling information management into the future. Security means that our systems must be secure from any and all enemy attacks. Trust implies that the information provided by any and all of our systems must be trusted, which means believable in all its forms. And finally, risk covers both the assessment of risk as well as its mitigation.

The National Institute of Standards and Technology (NIST) agency has explored a ransomware risk management profile [6]. Ransomware is a malicious attack in which the attackers encrypt an organization's information and data and then demand payment in return for not disclosing this information to a third party. The Profile in question identifies security objectives that deal with preventing, responding to, and recovering from ransomware events. This profile is a means to achieve risk management. It is envisioned to be used to help an organization in dealing with threats and the potential consequences of threat events.

DISA is also knee-deep in what they call a zero trust architecture [7]. This new architecture is expected to improve security, reduce complexity, and save money as the agency moves from the defense-in-depth approach to network security. The zero trust approach requires that every user and every device be authenticated before having access to the network. This approach also incorporates the following "systems":

a. Secure access service edge (SASE)
b. Software-defined area network (SD-WAN)
c. Identity credential access management (ICAM), and
d. Virtual security stacks

DECISION SUPPORT FRAMEWORKS

This is an area in which human decisions and software support come together. Future decisions will be aided by these software systems, which can also have the name "model-based systems engineering". In this regard, below is a list of some promising software system [8] in this category:

• Studio 5000
• AMESim
• System Link
• Enterprise Architect
• Altair Model-Based Development Suite
• Genesys (Vitech)
• Wolfram System Modeler
• Cameo Systems Modeler
• Cradle
• IBM Engineering Lifecycle Optimization
• Innoslate.

If we move on to software that is definitely associated with decisions, we obtain the following list [9]

- LogicNets
- 1000 minds
- Airfocus
- Collective [i]
- PICC Software
- Intuendi
- Checkbox
- Riskturn
- Statgraphics Centurion
- Style Intelligence
- Decisions
- Axiom Software
- Minitab
- Zingtree

Note that there is no software that is common to both lists. Or, in other words, the MBSE list does not recognize the "decision" capability as a feature.

BIG DATA AND ANALYTICS

Big Data

Big Data [10] is being acknowledged as important, day by day. Each and every day, big data is building up and ready for the analysts to explore its content and meaning. The systems engineer now needs to join these forces and consider how to deal with big data and how these data contribute to the issues that the systems engineer addresses. Big data is such that traditional software is not sufficient to deal with it.

Six concepts can be readily associated with big data:

1. Volume
2. Variety
3. Velocity
4. Veracity
5. Value
6. Variability

Volume, of course, refers to how much data there is. The size, as of today, is usually more than terabytes and well into petabytes. Variety deals with the type and nature of the data. Velocity refers to the rate at which the data is being generated and processed. Veracity deals with the truthfulness and reliability of the data. Value is the same as the overall worth of the data. And finally, variability considers the extent to which the data is structured or unstructured.

Analytics

It is a relatively new field of study and investigation, and has developed and grown in the past ten years or so [11]. It is relevant, in general, when one considers that it is the preferred method for dealing with big data. As an example of its scope, we present here certain aspects of an analytics course of study from the Wharton school.

The overall subject is built upon three notions:

1. Descriptive analytics
2. Predictive analytics
3. Prescriptive analytics

The Wharton school identifies program topics as below:

a. Gathering insights through descriptive methods
b. Being able to forecast
c. Making predictions using predictive methods
d. Use of analytics for decision-making
e. Making recommendation for change
f. Assessing most favorable outcomes

A week-by-week statement of course modules from Wharton is cited below:

Week 1 – Descriptive Analytics: gaining insights
Week 2 – Describing and forecasting future events
Week 3 – Making predictions using data
Week 4 – Application and toolkit
Week 5 – Tools for decision-making
Week 6 – Using data to predict employee performance
Week 7 – Recommendations for changing behavior
Week 8 – Determining most favorable outcomes
Week 9 – Applications for business

We make some additional observations in that (a) the above course is business-oriented, but can (and should) be mastered by tomorrow's systems engineers (b) there are specific tools for prediction and to support decision-making, (c) we believe we are able to predict employee performance (shades of George Orwell), and (d) the methods can be used to change behavior (more shades of Orwell).

This last chapter on "long poles in the tent" provided insights into five fields that need to be examined and mastered by the systems engineer a couple of decades down the road. The reader can see how this conception of the skills of tomorrow's systems engineer is considerable broader than that of today's. This is necessary in order for the future systems engineer to take his or her appropriate place in society, with the appropriate power, credibility, influence, and responsibility.

REFERENCES

1. See https://ac.cto.mil/digital_engineering
2. See https://sercuarc.org
3. "Digital Engineerng +Strategy", DoD, Systems Engineering, 2018.
4. INCOSE – Vision 2035, see incose.org
5. DISA, see www.disa.gov; also Ackerman, R., "Complementary Actions Define New DISA Strategy", SIGNAL, October 2021.
6. Barker, W. C., et al., "The Cybersecurity Framework Profile for Ransomware Risk Management", NIST, September 2021.
7. Seffers, G., "Thunderdome moves DISA Beyond Defense in Depth", SIGNAL, October 2021.
8. Model-Based Systems Engineering.
9. Decision Support Software.
10. See https://en.wikipedia.org
11. See Wharton Executive Education, Business Analytics: From Data to Insights, Online certificate program.

Miscellany

9

This chapter deals with a variety of miscellaneous topics, as follows:

- The acquisition innovation research center
- Bell Labs
- The Lemelson Center
- The book from Isaacson, with the title "The Innovators"
- The most innovative companies
- The most innovative universities
- Firms that have failed to innovate
- A citation of five ways to innovate and improving national innovation systems
- Artificial intelligence
- The Joint Artificial Intelligence Center (JAIC)
- Defense Advanced Research Projects Agency (DARPA)
- Systems Engineering Research Center (SERC)
- The System Dynamics Society
- Society Software
- Systems of Systems
- Rapid computer-aided system of systems engineering (RCASSE)
- The National Aviation System (NAS) Model
- Views of systems engineering outside the United States
- The future of systems engineering

This chapter cites various miscellaneous sources and information that appeared to not fit easily into the previous chapters, but is relevant to the overall subject.

DOI: 10.1201/9781003269274-9

INNOVATION

Not too long ago, I had occasion to watch a YouTube interview of William Perry, our Secretary of Defense from 1994 to 1997 [1]. Among the stories he told was one in which he was Director of the electronic defense labs of Sylvania, where he realized that a technology was coming along that very likely would put this part of Sylvania out of business, which was big into systems with vacuum tubes, and vacuum tubes themselves. Cash cow, one day; out of business the next. When he became SecDef, it was part of his charter to deal with this game changer, transistors replacing vacuum tubes everywhere. More about the transistor later, when we cite Shockley and his colleagues.

And there were other game changers; other technologies that made a difference. These included the Atom Bomb, Stealth Aircraft, the Home Computer, the chip replacing film, and others. All of these made big differences and represented game changers. And here we sit, looking at the future, and wondering what technologies are coming out, and what they will mean to us and our way of life.

Another perspective on innovation has come out of the Secretary of Defense, namely, from Kristen Baldwin in the Systems Engineering Office [2]. She was explicitly supporting innovation as set forth by the new SecDef, General Mattis. Three priorities were expressed:

a. More lethal force
b. Strengthening alliances
c. Achieving greater performance and affordability

A critical way that these are to be achieved is through innovation, which in turn comes about through new technology. This new technology is supported by major amounts of funding, and very careful technology insertion. Go back and think about the technology that Bill Perry talked about, namely, replacing vacuum tubes with transistors. Think also about the other technologies that are out there, from advanced miniaturized sensors, to high efficiency solar cells, and imagine where we might be (and need to be), a couple of decades down the road.

The Acquisition Innovation Research Center. This enterprise was established in 2020 with the following focus [3]:

> To infuse innovation and alternative methods needed to better respond to the rapid increase of technological advancements critical to today's warfighter.

Four major modes of behavior for the Center are:

a. Incubate
b. Ideate
c. Respond
d. Connect

Some of the activities of the Center include (a) establishing pilot programs to demonstrate acquisition insights, (b) enhance education efforts, (c) explore research efforts related to best practices, and (d) look for new leaders and professionals in this domain.

Bell Labs was the preeminent center of research in what was called the great age of American innovation [4]. This author can remember, back in 1957 when he graduated with a Bachelor's degree in electrical engineering, it was considered a great honor to be interviewed for a job at Bell Labs, and beyond that, being offered a job was the height of achievement. Several of my classmates were in the latter position.

The director of the Bell Labs for many years was not very well known, Merwin Kelly, who created a culture of innovation at the Labs. We cite here just a few of the "stories" that have come out of the Labs. Probably the most creative came about Claude Shannon, working in a modest office, and traversing the halls with a monocycle while trying to juggle. Shannon was a genius, and also somewhat of an eccentric. He created the field of information theory, just about single handedly. Of this activity, Bob Lucky, an executive at the Labs and also one of Shannon's disciples, commented that "I know of no greater work of genius in the annals of technological thought", as he referred to the work of Shannon. This author also appreciated the substance of information theory as he taught an introductory course in that domain at the George Washington University in the 60s.

Another "story" has been written about a researcher by the name of Shockley who, with two other researchers (Bardeen and Britain), won a Nobel Prize for their work in creating transistors. Not too far down that road, this new device replaced vacuum tubes in basically all electronic systems. This revolution into multifunctional chips literally transformed that entire industry. Yet another success story is that of John Pierce who led the charge at the Labs into communication engineering, including communication satellites. His early Echo balloon experiments are a classic example of the contributions of Pierce. Pierce tried hard to not seek recognition for his work, but rather claimed that his great skill was "to get others to do things". Perhaps the bottom line with respect to the Labs is that it was Merwin Kelly that "got others to do things".

The Lemelson Center, part of a museum that in turn has been a part of the Smithsonian, has been a leader in the study of invention and innovation [5].

They are housed in the Washington National Museum of Natural History from where they execute their charter to encourage Impact Inventing with the following three needs from an invention:

    a. It should have a positive social impact
    b. It should be naturally environmentally responsible
    c. It should be financially self-sustaining

They support invention and innovation largely by sponsoring projects with worthy researchers that propose new and interesting contributions. The Center taps into the strong national interest in innovation largely by providing funding and administrative support. This support is very much "hands-on" so they get a lot of credit for successful outcomes. It appears that interest across the United States has increased moving onto the 2000s, judging by the sheer number of projects as well as the new entities created inside the Universities.

    The Innovators [6]. A book from Walter Issacson has detailed the work and contributions made by many known to him as the Innovators. This author suggests that the systems engineer of tomorrow have the skill and the inclination to be a serious innovator, and also serve as a mentor for other innovators. Below is a list of a dozen innovators that are worth thinking about in this connection:

    1. Alan Turing. A genius from the UK who made important contributions to computation and cryptanalysis, including an important role in code breaking at Bletchley Park, UK
    2. Claude Shannon. As noted earlier in this chapter, established the basic foundations of information theory
    3. J. Presper Eckert and John Mauchly. Built an early computer at the University of Penn; then founded Eckert Mauchly Computer Corporation (in 1947) focused on government sales, initially to the Census Bureau and the Army Signal Corps
    4. John von Neumann. Another genius who made seminal contributions to many fields, including the foundations of mathematics, functional analysis, topology, computing, group and ergodic theory, and others
    5. Chester Carlson. Growing up in poverty in Seattle, he patented his ideas as to how to do electrophotography, leading to an entire industry of copying machines, including Xerox
    6. Marc Andreeseen. Developed the mosaic browser, and proceeded to support other technologists and technologies
    7. Vint Cerf and Bob Kahn. Important players in the Internet World, along with Bob Taylor and Tim Berners-Lee. Cerf had the position

of VP and Chief Internet Evangelist at Google, and Kahn was CEO for National Research Initiatives (CNRI)

8. Bill Gates. Lead person who built Microsoft, and made just the right moves with operating systems as well as home computer applications such as Microsoft Office

9. Larry Page and Sergey Brin. The two geniuses behind the best search engine known as Google; extended the state-of-the-art in this domain

10. Steven Jobs. Not an engineer but a most serious system designer; built Apple to one of our most successful companies

11. Andy Grove. Figured out how to build and provide chips for computers on a large scale so as to be the principal supplier of same

12. William Shockley. Along with Bardeen and Brattain, developed the transistor that replaced just about all vacuum tubes and led to the integrated circuit (chip), for which they won the Nobel Prize

MOST INNOVATIVE COMPANIES: 2021 AND 2020 [7]

Below is a comparison of the 2021 and 2020 ratings of most innovative companies and Universities.

Companies

2021	2020 (FORDHAM SURVEY)
1. Apple	1. Apple
2. Alphabet	2. Honda
3. Amazon	3. Weber
4. Microsoft	4. Toyota
5. Tesla	5. Amazon
6. Samsung	6. IKEA
7. IBM	7. Google
8. Huawei	8. Netflix
9. SONY	9. Navy Federal Credit Union
10. Pfizer	10. Samsung

Universities

2021	2020
1. Arizona State	1. Arizona State
2. Georgia State	2. Georgia State
3. MIT	3. MIT
4. Georgia Institute of Technology	4. Georgia Institute
5. Carnegie Mellon	5. Stanford
6. UMBC	6. Purdue
7. Stanford	7. Carnegie Mellon
8. Purdue	8. CalTech
9. Elon University	9. Northeastern
10. University of Michigan – Ann Arbor	10. UMBC

By way of further comparison, the innovation lists for the year before are shown below:

INNOVATIVE COLLEGES – YEAR BEFORE	INNOVATIVE COMPANIES – YEAR BEFORE
1. Arizona State	1. Apple
2. Stanford	2. Google
3. MIT	3. Microsoft
4. Georgia State	4. Amazon
5. Carnegie Mello	5. Samsung
6. Northeastern	6. Tesla
7. UMBC	7. Facebook
8. University of Michigan	8. IBM
9. Harvard	9. Uber
10. Duke	10. Alibaba

If you are thinking about sending an offspring to college and are looking for innovation, consider the US News and World Report that lists the following in positions 11 through 20 [8]:

1. CalTech
2. Johns Hopkins
3. Northeastern
4. University of California – Berkeley

5. University of Central Florida
6. Duke
7. Rice
8. George Mason University
9. University of Texas – Austin
10. Yale

FAILED TO INNOVATE [9]

Yet another look at the matter of innovation is the identification of some ten firms that failed to innovate. It is no surprise that most of them no longer exist:

1. Blockbuster
2. Polaroid
3. Toys R Us
4. Pan Am
5. Borders
6. PETS[DOT]COM
7. Tower Records
8. COMPAQ
9. General Motors
10. KODAK

Many of these enterprises were, at one time, considered leaders in their domains. What happened? They failed to "keep up" as their competitors moved past them.

A FEW WAYS TO INNOVATE

Let's suppose you have a cash cow product and you want to make sure it lasts as long as possible. Yes – you undertake an innovation program to try and protect that product for as long as possible.

a. Institute an R & D program "around" that product. You're looking for product improvements in performance. You're also trying to improve the "look and feel" of your product to make it more desirable for customers to use.

b. Increase the funding for a current R & D program. This allows you to bring new people and ideas to the table. New ideas include lateral thinking and disruptive thinking and Design thinking. Add a third sock to the pair. Put an elevator and plumbing outside the building. Review your patents for possible challengers. Buy your competitors' products and tear them apart.

c. Bring in to the company a sampling of customers and run some experiments with your product and their thoughts.

d. Study the market and especially your competitors. These give you some idea as to what directions to explore. Perhaps the answer lies in a completely different approach, such as a multi-functional product, or an automated product. Try to give your customer more "bang for the buck" in your next version of the product.

e. Build a new and stronger innovation culture, such that all employees are given incentives for innovating. Then don't forget to listen to them when they come up with new ideas. Along these lines, the author is reminded of a story about Xerox Parc, which goes as follows:

Xerox had taken a strong and progressive position by establishing Xerox Parc and providing outstanding funding for it (as well as Alan Kay as its director). However, as the story goes, when it reported progress and a need for new investments, the headquarters folks turned them down, ultimately in favor of putting new investments into real estate. This was both self-defeating and contrary to the original notion as to how to build a high-tech company. Soon the innovative people got the message and moved on to other enterprises, and the inevitable happened. The company began to slide downward and Xerox Parc lost its luster as well as many of its key people. It took the hiring of Anne Mulcahy to fix the problem. When she arrived, the company was 17 billion dollars in debt and a stock price that dropped from $63.69 a share to $4.43. Some three years later, Ms. Mulcahy had achieved four straight profitable quarters. Fortune Magazine called this performance "the hottest turnaround act since Lou Gerstner (of IBM)".

The reader is also referred to the author's text on problem-solving [7] in which there is a sampling of annual reports from about a dozen high-tech companies. Invariably, these companies indicate that innovation is important to them and their overall approach to their businesses. A typical phrase along these lines is simply "we view innovation as the lifeblood of our corporation".

IMPROVING NATIONAL INNOVATION SYSTEMS [10]

A mechanism for improving national innovation systems has been devised, under United Nations Conference on Trade and Development (UNCTAD), part of the U.N. This includes reviews called Science, Technology, and Innovation Policy (STIP) reviews. They address and diagnose the systems and raise various discussions regarding how to make improvements. They also increase awareness of problem areas so that systems can be properly adjusted. This mechanism is considered to be a valuable asset that can be brought into play in this arena.

ARTIFICIAL INTELLIGENCE

The SEBoK [11], provided by the SERC at the Stevens Institute of Technology, explored and documented the topic of artificial intelligence, making the following main points.

There is considerable attention given to "machine learning (ML)". The claim is that there are essentially three categories of learning:

1. Supervised
2. Unsupervised
3. Reinforcement learning

Further, the SEBoK makes the connection to the digital transformation, which "is the confluence of big data, cloud computing, AI and the Internet of Things (IoT)". The SEBoK envisions an enormous push into AI in relation to systems engineering and digital engineering, with AI applications, for example, into the following:

- Databases
- Cloud Storage
- ML learning frameworks
- Programming languages
- Data visualization
- Data and systems integration

The latter will be facilitated by advanced architectures and APIs.

THE JOINT ARTIFICIAL INTELLIGENCE CENTER [12]

The JAIC is considered transformative in the DoD. It states that it will transform joint warfighting and departmental processes by means of integrating AI, as well as bottoms-up results from DoD innovators. It also claims a holistic approach that addresses the following areas:

- The delivery and adoption of AI
- Defending against cyber attacks
- Evolving partnerships with industry and academia
- An AI workforce
- Military ethics and safety

The JAIC plans to delivery AI capabilities as national mission initiatives (NMIs) and component mission initiatives (CMIs). The former are broad and cross-cutting whereas the latter are more specific and intended to solve a particular problem.

AIS TURBULENT PAST [13]

An article on AI of special importance appeared in the October 2021 issue of the IEEE Spectrum [13]. It covered some of the history of AI (from 1956), pointing out that there are two major camps – the symbolists and the connectionists. The former were championed by the likes of Alan Sewell and Herbert Simon, and they generally supported the notion of rule-based expert systems. The latter have had many champions, more-or-less starting with Frank Rosenblatt at Cornell Labs who came up with the neural network approach that he called the perceptron. Advanced work by several connectionists has led to deep learning and applications in optical character recognition. Some successful researchers have found a way to integrate both approaches, which has been called a hybrid solution.

The author of the IEEE article suggests that AI is now in an upward cycle, with near record sponsorships. The author of this book is watching and waiting, looking for more evidence but with a clear bias in the direction of rule-based expert systems. We hope to see further advances in this area by tomorrow's systems engineering community. There is clearly a lot of good history behind this approach, and large numbers of application areas, such medical diagnostics and engineering trouble-shooting.

This same issue of the IEEE Spectrum featured an article addressing "ways AI fails" [14]. Here are the seven ways that AI can and does fail:

1. *Brittleness.* It is easily fooled by patterns it has not seen before, even though the new pattern is a simple rotation of an old pattern that it has seen before.
2. *Embedded Bias.* Biases present in the data in which the AI is trained are not eliminated; they show up in the AI decisions that are recommended.
3. *Catastrophic Forgetting.* AI trained to detect deepfakes do well with these, up to a point. But when confronted with new deepfakes, AI tends to forget the original deepfakes. A deepfake may, for example, be a picture of a celebrity.
4. *Explainability.* This is the inability of some AI to explain how they reached certain conclusions.
5. *Quantifying Uncertainty.* The lack of a method, for much of AI, to come to a number (or set of numbers) that represent levels of uncertainty.
6. *Common Sense.* Much of AI, in terms of its conclusions, lacks common sense that the human comes to quickly and intuitively.
7. *Math.* The researchers tell us – much of AI is simply not good at even easy math constructs.

Will the above "achilles heels" of AI impede its progress? Probably yes, but current evidence is that those devoted to embracing AI will continue to do so, and likely at ever increasing rates. There are just too many advantages (as per in robots, auto driving) to walk away from this reality as well as its potential for the future

AI SYSTEMS SOFTWARE [15]

A search for AI systems software revealed the following such systems, captured by an enterprise by the name of Capterra:

a. Solcure
b. Accern
c. Collective[i]
d. Genesys DX

e. Augmentir
f. Zero Incident Framework
g. Sisense
h. IpiphonyIntrohive
i. Answer Rocket
j. Introhive

The names certainly do not suggest that they are in the AI domain. It will take the next step of investigation to discover that fact. But if we move on into a sub-set of AI software known as expert systems, we obtain the following list of software packages [16]:

a. ES/P Advisor
b. Expert-Ease
c. Exsys
d. GURU
e. INSIGHT 1 and 2
f. KDS
g. KES
h. M.1
i. MICRO-PS
j. Personal Computer Plus
k. SeRIES-PC
l. TIMM-PC

Here again, the reader will need to take the next step of discovery to find out if a particular software package is applicable to the problem they are dealing with.

DARPA

This DARPA [17] has been in existence since 1958 and lays claim to some rather important technologies and systems. The latter have included:

a. The internet
b. Aspects of weather satellites
c. Parts of the global positioning system (GPS)
d. Drones
e. Stealth systems
f. Voice interfaces
g. The computer

The primary customer of DARPA, of course, is the Department of Defense. However, other sectors have benefitted, as in the example of the Internet and the GPS. Projects of interest tend to be high risk, high payoff in nature, and beyond standard short term approaches.

The agency is basically organized by Offices, as below:

- The Adaptive Execution Office
- The Defense Science Office
- The Information Innovation Office
- The Microsystems Technology Office
- The Strategic Technology Office
- The Tactical Technology Office
- The Biological Technologies Office

The reader who wishes to know more about the agency can consult its website at darpa.dod. Some of the more active research areas are cited below:

a. An anti-submarine warfare (ASW) vessel
b. Advanced jet fighters
c. Total airspace awareness
d. Atmospheric water extraction
e. Cancer research
f. Military satellite constellations
g. Military communications

It is of special note that DARPA has also sponsored work by Jet Propulsion Lab (JPL) to develop MBSE tools. These tools are to be used to perform business case analysis. This should be of particular interest to the systems engineer that is innovative and looking to develop an enterprise competing in the business world.

OTHER SERC [18]

We recall that the SERC is a systems research center at the Stevens Institute of Technology, sponsored by the systems engineering office in the Office of the Secretary of Defense (OSD). Important programs and projects at this time can be found under the following main topics:

1. Enterprises and System of Systems (ESOS)
2. Trusted Systems (TS)
3. Systems Engineering Management Transformation (SEMT)

4. Human Capital Development (HCD)
5. Research Incubators (RI)

Selected individual projects under the above include the ten projects listed below:

a. Mission engineering
b. Missiles and space systems engineering
c. SoS analytic workbench
d. Approaches to achieve modularity
e. Digital engineering
f. Systems engineering and systems management transformation
g. Systems engineering for velocity and agility
h. Systems engineering and analytics
i. Trusted systems
j. Research incubators

Looking at the SERC November 2021 meeting, we see a variety of research areas as well as four tutorials being presented. Both are cited below:

Tutorials (1) a MBSE Cost Model, (2) systems and cyber resilience model, (3) digital engineering, and (4) security engineering.

Research Reports:

- Acquisition Innovation
- Mission Engineering
- Digital Engineering
- AI and Autonomy
- Velocity
- Security
- Human capital development

It is observed that the SERC interests range far and wide, with excellent investigators. This work should definitely be reflected in the scope of systems engineering down the road.

SYSTEM DYNAMICS SOCIETY AND SOFTWARE [19]

The world of modeling and analysis was greatly enhanced by the work of Jay Forrester and his system dynamics model and methodology. We take special note of it here due to its overall utility and general contribution to the state-of-the-art. We have every reason to believe this field has a long life and a place in the minds (and hearts) of tomorrow's systems engineers.

The System Dynamics Society cites commonly used software in that domain as follows:

1. Dynamo
2. iThink
3. STELLA
4. Powersim Studio
5. Versim

The Society also claims that the following are extending the system dynamics methodology:

1. Anylogic
2. Dynanplan Smia
3. Goldsim
4. Berkley Madonna
5. Simile
6. Simgua
7. TRUE
8. Ventity

We see that there is no dearth of activity in this modeling and simulation world.

SYSTEMS OF SYSTEMS [20]

The SERC research indicates that the subject of systems of systems (SoS) is alive and well as of the writing of this book. This author believes that this topic is strong, and will remain strong for the indefinite future. Reason: systems are getting larger and more complex despite our efforts to simplify. An excellent example of a system of systems is the NAS. An overview suggests that the NAS has component systems such as:

a. Communication systems
b. Landing systems
c. Navigation systems
d. Personnel systems

e. Airport systems
f. Airspace systems
g. Air traffic control (ATC) systems
h. Support Software systems

Some years ago, this author carried out a program of research and formulated a concept relevant to this subject of systems of systems, which was documented under the category of RCASSE [21]. The ten elements of RCASSE were identified as:

a. Mission engineering
b. Baseline architecting
c. Performance assessment
d. Specialty engineering
e. Interface/compatibility evaluation
f. Software issues/sizing
g. Risk definition/mitigation
h. Scheduling
i. Preplanned product improvement
j. Life cycle cost-issue assessment

The basic idea behind RCASSE was that it would be executed within a nominal six-month period so that answers could be obtained quickly and therefore changes could be made before the various sub-system designs would be set in concrete. Today, this might draw some equivalence to agile system of systems engineering.

Research in the SoS arena tends to explain and explore the interactions between the subordinate systems, with an eye toward improvement and even optimization. This is sometimes given the name "synergy". A particularly cogent book on SoS and innovations [22] examines a variety of application areas, including:

a. openness
b. architecture
c. M & S
d. Net centricity
e. Integration
f. Emergence
g. Management
h. Defense
i. Health care
j. Space

A MODEL OF MODELS

A close relative of system of systems engineering is that of constructing a model of models. This was discovered and investigated when this author had a contract with the FAA to develop a NAS Model [23]. Some thirteen model areas were identified, as follows:

1. ATC Availability
2. Airport Capacity
3. Airport Delay
4. Airspace Capacity
5. Airspace Delay
6. Trip Time
7. Energy Utilization
8. Service Availability
9. Noise
10. Pollution
11. Security
12. Safety
13. Costs

The idea was (and is) that one starts, for each model, with the enumeration of the key dependent and independent variables. From there, one constructs an overview "model of models" that is used to study just about anything of importance in this domain. This model of models notion has not as yet received the attention that it warrants.

OVERSEAS VIEWS OF SYSTEMS ENGINEERING [24]

It is recognized that there are many interesting and forward-looking views of systems engineering that have originated outside the United States. Six researchers are cited, with apologies to those not documented here:

Joe Kasser. Has investigated "a unified theory of systems engineering" and explored systems thinking as it applies to systems engineering.

Derek Hitchins. Has examined and articulated a set of general principles applicable to systems.

H. Sillitto. Headed a group investigating a comprehensive approach to defining systems; wrote an important book on systems engineering.

Dov Dori. Known for his development of the Object Process Methodology, for which he received two important awards.

Olivier de Weck. Is credited with developing several systems engineering tools and methods, including the isoperformance method, spacenet simulation software, and the delta design structure matrix.

Heinz Stoewer. Served as President of INCOSE; contributed to space systems as managing director of the German Space Agency.

THE FUTURE OF SYSTEMS ENGINEERING [25]

We stop here for a moment to explore what Dr. Steven Dam has said about the future of systems engineering. First, some facts. Dr. Dam is the president and founder of SPEC Innovations, and one of the products of this firm is Innoslate, a model-based systems engineering tool. Thus, we would expect Dr. Dam to be a strong supporter of MBSE, and such is certainly the case. Here are some key points that Dr. Dam has made.

1. The field needs to embrace MBSE, and that likely starts with understanding of the Life-cycle modeling language (LML).
2. Given (1) above, the industry needs to understand and embrace the evolution of the LML to SySMl and more advanced versions of MBSE.
3. MBSE is the key to the evolution of digital engineering.
4. MBSE is the basis for the integration of tools.
5. MBSE tackles one of the more serious systems problems, namely, that of increasing complexity (e.g., a Zettabyte of data – 10th to the 21st).
6. Innoslate is an important instantiation of MBSE, and practitioners of systems engineering should keep that in mind.
7. Looking to the future, we see the systems engineer as fully versed in MBSE, both in terms of development and also in terms of use.

INCOSE'S VIEW OF THE FUTURE [26]

Although INCOSE's view of the future is addressed elsewhere in this treatise, some additional points are considered as this chapter is brought to an end.

The state-of-the-art of systems engineering will evolve, in the next couple of decades, to create new and more stringent stakeholder expectations. So whatever the stakeholder expects today, that is likely to increase dramatically as we move on down the road. The systems engineer must be prepared for this increase, or if not, may well lose ground in the continuing race for relevancy.

The evolution of systems engineering will also place new demands upon the workforce. The latter needs to have greater skills and more powerful systems engineering tools to bring to the table. This points to MBSE and also to digital engineering. It also creates various obligations on the part of systems engineering management as well as academia. The latter needs to provide leading edge ideas and methods in order to stay relevant and be a driving force for change.

And finally, the reader should take time out to read INCOSE's view of the future, and figure out where he or she fits into that future. The bottom line is that the systems engineer will make that future, and also that the future will make the systems engineer as well as the field of systems engineering at large.

POTENTIAL FUTURE ACTIONS

In this section, the author suggests some future actions that might be taken by INCOSE, or the OSD or the SERC. Basically, the idea is that these stakeholders form a committee to investigate this book's contents (from a prior NASA executive to a current set of comment by an INCOSE Fellow), and come to some independent conclusions and actions.

REFERENCES

1. William, P., Secretary of Defense, "Interview by Stanford Student Group", YouTube.
2. Baldwin, K., "DoD Systems Engineering for National Security", IS 2018, INCOSE YouTube.

3. The Acquisition Innovation Research Center.
4. Gertner, J., "The Idea Factory", Penguin Books, 2012.
5. The Lemelson Center, see https://en.wiki
6. Issacson, W., "The Innovators", Simon & Schuster, 2014.
7. Eisner, H., "Problem-Solving", CRC Press, 2021; see p. 102 (Innovation).
8. US News and World Report.
9. Glaveski, S., "Collective Campus: Failed to Innovate".
10. UNCTAD (United Nations Conference on Trade and Development).
11. SEBoK, "Artificial Intelligence", SERC, Stevens Institute of Technology, see https://www.sebokwiki.org
12. JAIC, Joint Artificial Intelligence Center.
13. Strickland, E., "The Turbulent Past and Uncertain Future of AI", IEEE Spectrum, October 2021.
14. Choi, C., "Seven Revealing Ways AI Fails", IEEE Spectrum, October 2021.
15. AI Systems Software, Capterra.
16. Eisner, H., "Problem-Solving", CRC Press, 2021, see pp. 84–86.
17. DARPA – Defense Advanced Research Projects Agency.
18. SERC – Systems Engineering Research Center.
19. System Dynamics Society and Modeling.
20. SoS.
21. RCASSE.
22. Jamshidi, M., (Ed.), "Systems of Systems Engineering: Innovations for the 21st Century".
23. Models of Models.
24. Selected Views from Overseas.
25. The Future of SE (Dam); see https://specinnovations.com/dr_steven_dam; see also www.youtube.com
26. INCOSE Vision.

Summary

10

CHAPTER 1 – TOWARD A FUTURE ENVIRONMENT

The force fields that are relevant today, as they were some 20 years ago are:

a. The information age
b. Speed and responsiveness
c. Competition
d. New work patterns and environments
e. Loyalties and leverage

Two major stresses on the overall field of systems engineering were documented as:

1. Mike Griffin's position as exemplified by "how do we fix systems engineering?"
2. An INCOSE Fellow's position that we are doing "systems engineering management", and not hard core systems engineering

Three sources that tend to represent what systems engineering is all about are:

a. The INCOSE journal "systems engineering"
b. The systems engineering handbook (v. 4)
c. The INCOSE look at the field by the year 2035 (INCOSE vision)

The latter document cites systems engineering goals as:

a. To align systems engineering initiatives
b. To address future challenges

DOI: 10.1201/9781003269274-10

    c. The broaden the base of systems engineering practitioners
    d. To promote systems engineering research

Specific areas of interest at the OSD level include:

    a. Digital engineering
    b. The modular open systems approach (MOSA)
    c. Software
    d. Mission engineering/system of system engineering
    e. Sustainment engineering
    f. Trusted hardware and software assurance
    g. Cyber resilient weapon systems, and modeling and simulation

The top "baker's dozen" areas were cited as below:

1. Knowledge, data and information management
2. Interactive MBSE, including M & S
3. Agility, resilience, and adaptability in systems engineering procedures, processes, and practices
4. Establishing systems engineering as its own rigorous discipline
5. Security, trust and risk analysis, and mitigation
6. Affordability
7. SE development environments and tools
8. Handling the increased complexity of systems, and systems of systems (SoS)
9. More elegant design and architectures
10. Information technology (IT) system readiness
11. Integrated modular approaches
12. Software systems engineering
13. Digital engineering

CHAPTER 2 – TOMORROW'S SYSTEMS ENGINEER

This "super" systems engineer can aspire to the following:

    a. Grow into a "super" systems architect
    b. Be a Superior team leader on challenging projects
    c. Be a major contributor to the literature
    d. Be one step down from Corporate Vice President
    e. Be an entrepreneur

Success Factors for the systems engineer, in addition to the above, are:

1. Synthesizer
2. Listener
3. Curious systems thinker
4. Manager/Leader
5. Expert/ESEP
6. Expert/Domain Knowledge
7. Perseverer

Pure management skills of the super systems engineer are:

a. Building a high performance integrated team
b. Assuring that the system is built fast and more cheaply
c. Interfacing with stakeholders
d. Interfacing with top management and Congress

Three critical technical roles include:

1. Developing and assuring a superior system architecture
2. Achieving high technical performance from the system itself
3. Contributing to the field of systems engineering

Achieving levels of skill and performance comparable to some of our best systems engineers and managers, as for example:

- Eberhardt Rechtin
- Norman Augustine
- Simon Ramo
- Andy Grove
- Irwin Jacobs
- Andy Viterbi
- Robert Oppenheimer
- Admiral Hyman Rickover

CHAPTER 3 – KINDS AND TYPES BEYOND THE TIPPING POINT

Today's systems engineer must be good at synthesis.

Analysis is the companion of synthesis, but it must take second place.

We are seeking a new reality of "build once and use many times".

Consider Gladwell's very successful approach:

   a. Formulate a sound idea
   b. Articulate the idea and its derivatives/examples
   c. Use many chapters to explain and explore

Tomorrow's systems engineer needs to be broader and much more well-read. See the list of a dozen recommended books. Also, see the list of visionary companies, as below:

   1. American Express
   2. Boeing
   3. Ford
   4. General Electric
   5. Hewlett Packard
   6. IBM
   7. Marriott
   8. Walmart
   9. Disney
   10. Procter & Gamble

CHAPTER 4 – THE SYSTEMS APPROACH AND THINKING

The systems approach can be articulated by means of the following ten elements:

   1. A systematic and repeatable process
   2. Interoperability and harmonious system operation
   3. Consideration of alternatives
   4. Iterate to refine and converge
   5. Create a robust and slow-die system
   6. Satisfy all agreed-upon requirements
   7. Provide a cost-effective solution
   8. Assure the system's sustainability
   9. Utilize advanced technology
   10. Employ systems thinking

Six fundamental concepts for systems thinkers are:

   a. Interconnectedness
   b. Synthesis

c. Systems mapping
d. Feedback loops
e. Causality

"What business are we in", as part of strategic planning, often reveals the difference between systems and non-systems thinking
The successful systems engineer has the following associations:

1. A team player
2. Continuous learning
3. Creative
4. Problem-solving
5. Analytic ability
6. Communication skills
7. Logical thinking
8. Attention to detail
9. Mathematical ability
10. Leadership

Leading sources of information about systems thinking:

- Peter Senge
- The INCOSE Handbook
- Russ Ackoff
- Peter Checkland
- John Boardman and Brian Sauser
- Michael Jackson
- Donella Meadows
- Jamshid Gharajedaghi
- Gerald Weinberg
- S. G. Haines
- Virginia Anderson and Lauren Johnson
- Daniel Kim
- L. B. Sweeney
- Barry Richmond
- Michael McCurley
- John Gall
- O'Connor and McDermott
- Albert Rutherford
- Mmm Whitcomb
- Marcus Dawson
- Michael Goodman

- F. Capra and P. L. Luisi
- Monat and Gannon
- Rosalind Armson

CHAPTER 5 – THE SYSTEMS ENGINEER AS EXECUTIVE VP

Four management skills for this systems engineer have been cited as:

1. Building a high performance integrated team
2. Assuring that the system is built quickly and less expensively
3. Interfacing with stakeholders
4. Interfacing with top management and congress, if necessary

Three technical roles for this same systems engineer are:

a. Developing and assuring a superior system architecture
b. Achieving high technical performance from the system itself
c. Contributing to the field of systems engineering

Comments on Robert Oppenheimer as super systems engineer:

He showed a refined, sure informal touch.

His grasp of problems was immediate.

He could turn on – and turn off – his personal charm.

The systems engineer of tomorrow will be able to serve as the executive VP of any enterprise, including those with high technical content.

CHAPTER 6 – ARCHITECTING AND MODELING

Selected definitions of an architecture:

DoDAF: "The structure of components, their relationships, and the principles and guidelines governing their design and evolution over time".

This Author: "An organized top-down selection and description of design choices for all the system functions and sub-functions, placed in a context to ensure interoperability and satisfaction of final system requirements".

The NAS Model: An overarching model that deals with the following attributes of the NAS:

a. Airport capacity
b. Airspace capacity
c. Demand
d. Trip time
e. Delay
f. ATC Availability
g. Service Availability
h. Landing system performance
i. Pollution
j. Noise
k. Energy utilization
l. Safety
m. Security

Elements of the National Missile Defense program have been noted as:

1. Ground-based interceptor missiles
2. The Aegis ballistic missile defense system
3. Terminal high altitude area defense
4. Airborne systems

Simulation Software

Many software packages are available for use by the systems engineer. For the old timer, here are some of the software offerings:

a. SIMLAB
b. SIMAN
c. GPSS
d. DYNAMO
e. SIMSCRIPT
f. SLAM
g. GASP

Model-Based Systems Engineering (MBSE) is here to stay, and continue to be invested in over the next couple of decades. As such, MBSE packages will be made available off-the-shelf.

CHAPTER 7 – SOFTWARE REUSE AND AGILITY

Software reuse continues to show potential for major advances in "faster, cheaper, better". The DOTSS approach suggested improvements of 2,500% (!). A pointer toward reuse was seen as far back as 1968, from Bell Labs. Their comment was a proposed "basing of the software industry on reusable components". Will the industry step up to this challenge? Time will tell, but if it doesn't, a great opportunity will be lost.

Four core elements for agile acquisition have been identified as:

a. A program structure with small and frequent capability releases
b. Placing value on the software itself
c. Maximally responsive to changing of all types
d. Active involvement of users

Today's agility elements include releases, sprints and scrums.

Software architecting: The mysteries of software architecting will continue to exist, although practitioners will work hard to explain what they do and how they do it. This author suggests that his overall hardware procedure is adequate for inclusion into software architecting processes.

Resilience: We note Yakov Haimes' definition:

Resilience is the ability of a system to withstand a major disruption within acceptable degradation parameters and to recover within an acceptable time and composite costs and risks.

CHAPTER 8 – DATA AND DECISIONS

Considerable time and attention is being paid to "digital engineering" which appears to be on everyone's wish list for the future. Benefits in terms of this capability include:

a. Better transparency and decision-making
b. Greater adaptability
c. Increased efficiency

INCOSE has gone on record in their vision for the future with the following statement:

advancing digital technologies are enabling MBSE practices, but emerging computational capabilities, cloud infrastructure, and data information discovery are not being fully leveraged compared to other engineering and scientific disciplines.

CHAPTER 9 – MISCELLANY

Modes of Behavior for the Acquisition Innovation Research Center

a. Incubate
b. Ideate
c. Respond, and
d. Connect

Most Innovative Companies – 2021

1. Apple
2. Honda
3. Weber
4. Toyota
5. Amazon
6. IKEA
7. Google
8. Netflix
9. Navy Federal Credit Union
10. Samsung

Most Innovative Universities – 2021

1. Arizona State
2. Georgia State
3. MIT

4. Georgia Institute of Technology
5. Carnegie Mellon
6. UMBC
7. Stanford
8. Purdue
9. Elon University
10. University of Michigan – Ann Arbor

DARPA – Active Research Areas

a. ASW vessels
b. Advanced jet fighters
c. Total airspace awareness
d. Atmospheric water extraction
e. Cancer research
f. Military satellite constellations
g. Military communications

System Dynamics Society – commonly used software

1. Dynamo
2. iThink
3. STELLA
4. Powersim Studio
5. Versim

SERC Research Areas

a. Enterprises and System of Systems
b. Trusted Systems
c. Systems Engineering Management Transformation
d. Human Capital Development
e. Research Incubators

Models of National Aviation System (NAS) Model

1. ATC Availability
2. Airport Capacity
3. Airspace Capacity
4. Airport Delay
5. Trip Time

6. Energy Utilization
7. Service Availability
8. Noise
9. Pollution
10. Security
11. Safety
12. Costs

Short Form Suggestions for Investment for Tomorrow

Looking at all the current investments, and going beyond these, here is a short list of final suggestions for investment in the field of systems engineering:

- System architecting
- Software reuse
- MOSA
- Digital engineering
- Trusted software
- Cybersecurity
- Resilient and agile systems
- MBSE (Model-Based Systems Engineering)

Suggestions for Today's Systems Engineer

This section provides suggestions for today's systems engineer in terms of reaching for "super" status and performance a couple of decades down the road.

As a Member of a Company

a. Lead the charge, within the company, to obtain advanced certification and skills (i.e., Masters in systems engineering and ESEP status from INCOSE)
b. Start a community of practice (COP) in Systems Engineering
c. Have regular seminars (at least monthly) on special topics in systems engineering
d. Participate regularly in writing proposals to develop a research program
e. Contribute regularly by publishing in "systems engineering" and other journals
f. Assure continuous learning in systems engineering (as per Peter Senge)

As An Individual

    a. Read constantly about the latest practices in systems engineering
    b. Get to know the literature, and your favorite writer/researcher
    c. Write a book or two (or more)
    d. Meet with potential customers
    e. Attend webinars regularly

Index

Printed in the United States
by Baker & Taylor Publisher Services